OXFORD

GCSE

Higher

Homew

Clare Plass formerly of Trent College

About this book

This book has been written to
covered within the Higher Stuc
each Student Book unit: Home
Homework 2 covers unit lessor
lessons 3 and 4 and Homework
as reviewing the rest of the un

Contents

OXFORD
UNIVERSITY PRESS

OXFORD
UNIVERSITY PRESS

Great Clarendon Street, Oxford OX2 6DP

Oxford University Press is a department of the University of Oxford.
It furthers the University's objective of excellence in research, scholarship,
and education by publishing worldwide in

Oxford New York

Auckland Cape Town Dar es Salaam Hong Kong Karachi
Kuala Lumpur Madrid Melbourne Mexico City Nairobi
New Delhi Shanghai Taipei Toronto

With offices in

Argentina Austria Brazil Chile Czech Republic France Greece
Guatemala Hungary Italy Japan South Korea Poland Portugal
Singapore Switzerland Thailand Turkey Ukraine Vietnam

Oxford is a registered trade mark of Oxford University Press
in the UK and in certain other countries

© Appleton et al. 2006

The moral rights of the author have been asserted

Database right Oxford University Press (maker)

First published 2006

British Library Cataloguing in Publication Data

Data available

ISBN 978-0-19-915080-9

10 9 8 7 6

Typeset by MCS Publishing Services Ltd., Salisbury, Wiltshire

Printed in Great Britain by Ashford Colour Press Ltd., Gosport

Front cover photo: Image DJ

1 Work out these, giving your answer in its simplest form.

a $\frac{1}{4} + \frac{5}{8}$　　**b** $\frac{5}{7} - \frac{2}{5}$　　**c** $3\frac{2}{7} - \frac{5}{8}$　　**d** $\frac{23}{16} - \frac{-8}{15}$

e $\frac{4}{9} \times \frac{3}{4}$　　**f** $2\frac{1}{2} \div \frac{5}{9}$　　**g** $4\frac{1}{2} \times 1\frac{13}{27}$　　**h** $4\frac{2}{3} \div \frac{7}{8}$

2 Solve these equations.

a $4x - 1 = 15$　　　　**b** $9 = 3x + 7$

c $2x + 8 = 5x - 4$　　**d** $1 - x = 3x + 11$

e $\frac{x}{5} = 2$　　　　　**f** $4(2x + 5) = 34$

3 Use Pythagoras' theorem to find the missing side, x, giving your answer to 3 significant figures.

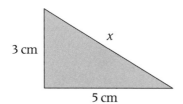

3 cm

x

5 cm

4 A bag contains red, blue and yellow balls.
The probability of selecting a red ball is 0.15.
The probability of selecting a blue ball is 0.4.

a What is the probability of selecting a yellow ball?
b What is the probability of selecting a ball that is not red?
c If there are 20 balls in the bag, how many blue balls are there?

1 Write these sets of numbers in ascending order.

  a 0.04, 0.14, 0.004, 4, 1.4

  b 3.92, 9.32, 3.29, 32.9, 0.329

2 Round these numbers to the nearest
   i 10 and
   ii 100.

  a 493
  b 207
  c 94
  d 1046
  e 17 320

3 Round these numbers to
   i 1 decimal place and
   ii 2 decimal places.

  a 138.345
  b 0.628
  c 4.529
  d 0.0994
  e 63.999

4 Calculate each of these.

  a 62.7×100 **b** $945 \div 10$
  c 2.679×100 **d** $3472 \div 1000$
  e 0.0043×100 **f** $35.29 \div 100$
  g 2.004×100 **h** $0.030\,45 \times 1000$

5 Work out these calculations.

  a $-8 - 3$ **b** $14 - -6$
  c $-3 + 10$ **d** $-9 + 7$
  e $-3 - 17$ **f** $-10 - -5$
  g $-4 + 100$ **h** $17 - -50$
  i $3 + -5 - -8$ **j** $19 - -4 + 8$

For questions 1 to 3, do not use a calculator.

1 Work out these calculations.

    a 2×-3 **b** -7×-9 **c** $-12 \div -3$
    d $24 \div -6$ **e** -5×12 **f** 9×-3
    g $-35 \div -7$ **h** $-56 \div 8$ **i** $10 \div -\frac{1}{2}$
    j $-16 \div -4 \times 6$

    Hint: Use BIDMAS in part **j**

2 Work out these calculations.

    a 3.6×-10 **b** -4.2×10 **c** $973 \div -100$
    d $-54.6 \div 10$ **e** 0.074×-1000 **f** $5.25 \div -100$

3 Calculate each of these.

    a $\dfrac{-8 \times -4}{16}$ **b** $\dfrac{-100 \div 4}{5 \times -5}$ **c** $\dfrac{(-4)^2}{56 \div -7}$

    d $\dfrac{15 - -25}{(-2)^3}$ **e** $\dfrac{-5 \div \frac{1}{3}}{-3 + 8}$ **f** $\dfrac{(-5 - 2)^2}{13 - 6}$

4 Use a factor tree to write each of these numbers as a product of its prime factors.

    a 60   **b** 210   **c** 378   **d** 504   **e** 2156

5 A cuboid is made from 330 small cubes. Using the prime factors of 330 to help you, list the dimensions of all cuboids that can be made with 330 small cubes.

    Hint: Some of the possible sets
        of dimensions contain a 1
        e.g. $1 \times 2 \times 165$.

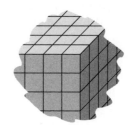

1 By drawing an appropriate Venn Diagram, find the
  i HCF and
  ii LCM
   of these pairs of numbers.

  a 36 and 60
  b 45 and 120
  c 54 and 504

2 Isla beats her toy drum every 5 seconds. Ayesha blows a whistle every 12 seconds.

If Isla and Ayesha begin at the same time, when do they next play their instruments together?

Hint: Find the LCM of 5 and 12. Explain why this works.

3 Work out $100 - 99 + 98 - 97 + ... + 4 - 3 + 2 - 1$

4 a and b are prime numbers.

$a^2 b^3 = 1125$

Find the values of a and b.

5 A pair of numbers that have a HCF of 1 are known as co-prime. Which of these pairs of numbers are co-prime?
  a 14 and 25
  b 12 and 21
  c 35 and 57
  d 42 and 105

1 Round these numbers to 1 d.p.

   a 3.618   **b** 12.406

   c 5.289   **d** 0.951

   e 100.95   **f** 1032.945

2 Work out these without using a calculator.

   a 6.3×10 **b** $4.92 \div 10$

   c 0.036×100 **d** $8.53 \div 1000$

   e $0.000\,86 \times 100$ **f** $10.02 \div 100$

3 Calculate each of these.

   a $6 + -3$ **b** -10×-6

   c $-3 \div 100$ **d** $28 - -9$

   e -3×-8 **f** 0.7×-100

   g $51 \div -3$ **h** $-3 + -12$

4 **a** Express 36 as the product of its prime factors.

   b Use a Venn diagram to find the HCF of 36 and 90.

   c What is the LCM of 36 and 90?

5 Draw a sketch diagram of each of these solids and state whether the solid is a prism or not a prism.

   a Cylinder
   b Cuboid
   c Tetrahedron
   d Square-based pyramid
   e Cone

1 For each of these shapes calculate
   i the area and
   ii the perimeter.

a

9 cm, 4 cm

b

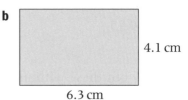

6.3 cm, 4.1 cm

c

13 cm, 5 cm, 12 cm

d

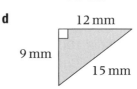

12 mm, 9 mm, 15 mm

2 Calculate the area of each shape.

a

12 cm, 3 cm, 9 cm, 8 cm

b

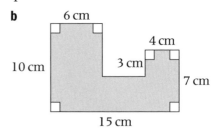

6 cm, 4 cm, 10 cm, 3 cm, 7 cm, 15 cm

3 Calculate the area of each shape.

a

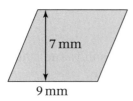

7 mm, 9 mm

b

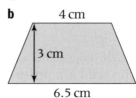

4 cm, 3 cm, 6.5 cm

4 ABCD is a trapezium where AB is parallel to DC.
   DC = 12 cm and AB : DC = 3 : 4.
   The distance between AB and DC is 5 cm.
   Calculate the area of the trapezium ABCD.

   HINT: Draw the shape first, labelling the vertices clockwise.

Example

Calculate the circumference and area of a circle, radius 4 cm.

Circumference = $2\pi r = 2 \times \pi \times 4 = 25.1$ cm (to 1 d.p.)
Area = $\pi r^2 = \pi \times 4^2 = 50.3$ cm^2 (to 1 d.p.)

1 Calculate the circumference of these circles.

a   4 cm

b   1.9 mm

c 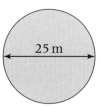   25 m

2 Calculate the areas of the circles in question 1.

3 Jason has a lawn which is a rectangle with a semi-circle at each end.
Calculate the area and perimeter of Jason's lawn.

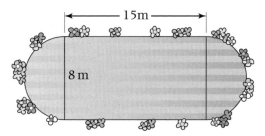

15m

8 m

4 A circular flowerbed of radius 1.4 m is surrounded by a path of width 70 cm.

Calculate the area of the path.

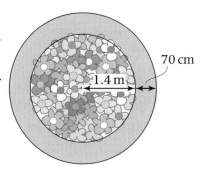

70 cm

1.4 m

1 Work out the surface area of these prisms, giving your answer to 3 significant figures where necessary.

a

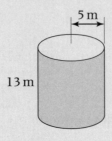

28 mm

90 mm

42 mm

b

8 cm   10 cm

6 cm   14 cm

c

5 m

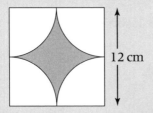

13 m

2 Calculate the shaded area.

12 cm

3 The area of this rectangle is twice the area of the square. Find the length of a side of the square.

4 m

18 m

4 A circle has an area of 98 mm^2.
Calculate the circumference of the circle.

Hint: Substitute 98 mm^2 into $A = \pi r^2$ and rearrange to find the radius.

1 Work out these questions on negative numbers.

a $-4 + 8$ b $4 - -3$ c -6×-5

d $12 \div -2$ e $-10 \div \frac{1}{3}$ f $-3 - 5 + 2$

g $4 \times -6 \div 2$ h $\dfrac{-3 \times 5}{-2 + 5}$ i $\dfrac{(-2)^3}{13 - -3}$

2 a i Express 60 and 108 as products of their prime factors.
  ii Use your answer to part **i** to work out the HCF of 60 and 108.

  b If $1.42 \times 230 = 326.6$, write the answer to

   i 14.2×23 **ii** 0.142×2.3

3 This is the base of a prism with vertical height 25 cm.

Calculate the volume of the prism.

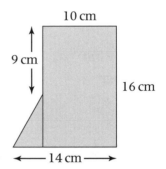

4 This frequency chart shows the heights of 30 children.
Copy and complete this frequency table for the information on the chart.

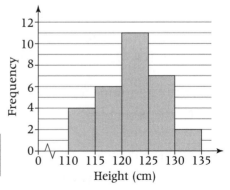

Height, h, cm	Frequency
$110 < h \leqslant 115$	

Example

a	Simplify this expression $5x - 3 + 2x$
b	Evaluate the expression, given $x = 5$.

a $5x - 3 + 2x = 5x + 2x - 3 = 7x - 3$
b $(7 \times 5) - 3 = 35 - 3 = 32$

1 Evaluate these, given that $x = 3$.

    a $2x + 7$ **b** $5 - x$ **c** $x^2 + 2x$

    d $\dfrac{3x - 4}{5}$ **e** $2x^2$ **f** $\dfrac{2x - 1}{6x - 3}$

2 Simplify these expressions, where possible.

    a $2 \times x$ **b** $y \times y \times y$

    c $4x + 9x$ **d** $5a + 3b$

    e $7m - n - 3m - 2n$ **f** $x^2 + 5x + 3x^2$

    g $3pq + 8qp$ **h** $2p \times 4q$

    i $\dfrac{9a}{a}$ **j** $\dfrac{30x^2}{6xy}$

3 Expand these brackets and simplify where possible.

    a $3(x + 4)$ **b** $y(y - 3)$

    c $-5(p + 2q - r)$ **d** $4m(m - n)$

    e $2(a + 3) + 5(b + 4)$ **f** $2(x + y) - 4(3x - y)$

    g $5(h + 2) - 4(2 - h)$ **h** $(p - q) - (2 - q)$

4 a Write a formula for

    i the perimeter
    ii the area

    of this rectangle.

  b If the area is 40 cm², show that

    $15x - 45 = 0$

5

$3x - 1$

1 Expand and simplify these expressions.

    a $(x+3)(x+4)$ **b** $(y+4)(y-1)$

    c $(3+p)^2$ **d** $(x+5)(x-5)$

    e $(h-5)(h+2)$ **f** $(2x+3)(x+5)$

    g $(3t+1)(2t+5)$ **h** $(3m+4)^2$

    i $(3p+q)(2p-3q)$ **j** $(4x-3y)^2$

2 If the area of the rectangle is 77 cm², show that

$$6x^2 - 17x - 65 = 0$$

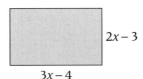

$2x-3$

$3x-4$

Factorise $2m^3 - 4m^2$

$2m^3 - 4m^2 = m^2(2m-4) = 2m^2(m-2)$

3 Factorise these expressions.

    a $5x+10$ **b** $3x-9$

    c $12x-2t$ **d** $4ab-2a$

    e $6xy+9x-12y^2$ **f** $8h^3-h$

4 Factorise these expressions.

    a $(a+b)+3(a+b)^2$ **b** $(p+qr)^2-6(p+qr)$

    c $pq+rq+px+rx$ **d** $xy+xw-2y-2w$

5 Use factorisation to help you evaluate these without using a calculator.

    a $3 \times 0.43 + 3 \times 1.57$

    b $5 \times 4.93 - 5 \times 2.73$

    c $4.78^2 + 4.78 \times 5.22$

1 Factorise these expressions into double brackets.

a $x^2 + 7x + 12$ **b** $x^2 + 6x - 16$

c $x^2 - 4x - 45$ **d** $x^2 - 11x + 28$

e $x^2 - 8x + 16$ **f** $x^2 - 49$

g $x^2 - 2x - 120$ **h** $x^2 - 196$

2 a Simplify these expressions.

    i $6p + 2p - 3q$

    ii $5a \times 3b$

   b Expand $4(5 - 3x)$

   c Expand and simplify $4(2y + 1) - 3(5y - 2)$

3 a Simplify $3a^2b \times 8a^3b^2$

   b Factorise completely $x^2 + xy - 2x - 2y$

4 The diagram shows a rectangle
with length $x + 5$ and width $x - 3$.
The perimeter of the rectangle is
P cm.
The area of the rectangle is A cm.

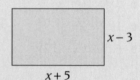

$x - 3$

$x + 5$

   a Show that $P = 4x + 4$.

   b Show that $A = x^2 + 2x - 15$.

5 a Expand and simplify $(p + q)^2$.

   b Hence or otherwise, find the value of

$$1.28^2 + 2 \times 1.28 \times 2.72 + 2.72^2$$

N2 HW1 Review

1 Work these out *without* using a calculator.

   a 2.7×100 **b** 0.34×10

   c $78.9 \div 10$ **d** 0.05×1000

   e $984 \div 100$ **f** $10.13 \div 10$

   g 42.05×1000 **h** $3.1 \div 1000$

2 Simplify these expressions, where possible.

   a $4x \times 3y$ **b** $2p \times 4p$

   c $a \times a \times 3a$ **d** $5ba + 3ab + bc$

   e $3m - n + 4n + 2m$ **f** $x^2 + 3x^3 - 2x^2$

   g $\dfrac{6gk}{3k^2}$ **h** $\dfrac{8y^4}{y^3}$

3 Calculate the area of this cross-section of Jacob's toy rocket.

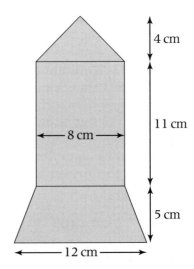

4 State the order of rotational symmetry of these shapes.

   a a square   Hint: It might help to draw a sketch

   b a regular pentagon

   c an isosceles triangle

   d a rectangle

1 Round these numbers to the degree of accuracy given in brackets.

a 5376 (nearest 1000)   b 67 (nearest 10)
c 63.2 (nearest whole)   d 24.25 (1 decimal place)
e 4289 (nearest 100)   f 15.282 (2 decimal places)
g 0.8758 (3 decimal places) h 1.96 (1 decimal place)

Example

Round 3.457 to 2 significant figures.

$3.457 \approx 3.5$

2 Round these numbers to the degree of accuracy given in brackets.

a 1.42 (2 significant figures)
b 25.331 (3 significant figures)
c 0.0321 (1 significant figure)
d 0.702 (2 significant figures)
e 5248 (3 significant figures)
f 42 735 (1 significant figure)
g 107.65 (4 significant figures)
h 95 950 (3 significant figures)

3 By rounding all of the numbers to 1 significant figure, estimate the answer to each of these calculations.

a 56×923

b $15\ 048 \div 456$

c $704 + 2520 \div 105$

d $\dfrac{52.3 - 12.8}{2.93 + 5.06}$

e $\dfrac{32.71 \times 4.09}{2.15^2}$

f $\dfrac{3.93 - 2.57}{12.5 \times 2.6}$

Hint: Don't forget to use BIDMAS!

4 Work out these questions mentally.

a $2.3 + 5.2$ b $12.5 + 0.7$ c $8.4 - 4.9$
d $4.56 - 3.5$ e $15 - 3.6$ f $6.52 + 3.48$
g $7.26 - 6.85$ h $12.4 - 9.68$

Do not use a calculator for these questions.

1 Work out each of these questions using written methods.

    a $5.64 + 3.2$ **b** $3.5 + 0.482$

    c $72.85 + 12.002$ **d** $7.85 - 5.93$

    e $24.35 - 18.9$ **f** $15.04 + 2.681$

    g $54.862 - 35.97$ **h** $104.25 - 68.489$

2 a Write the answer to 8×16.

    b Use your answer to part **a** to work out

       i 0.8×16 **ii** 8×0.16

       iii 80×0.016 **iv** 800×1.6

       v 0.008×0.016

3 Given that $8.32 \times 640 = 5324.8$, write the answer to

    a 83.2×64 **b** 0.832×6.4

    c $532.48 \div 6.4$ **d** 0.832×0.64

    e $53.248 \div 83.2$ **f** $5.3248 \div 0.64$

4 Work out these questions without using a calculator.

    a 0.4×0.3 **b** 5×0.8

    c $1.2 \div 0.04$ **d** $750 \div 1.5$

    e 43.5×0.2 **f** $0.96 \div 0.016$

    g $19.6 \div 0.14$ **h** 1.32×0.3

5 You can use one calculation to derive a family of facts.
For example, from $4 \times 9 = 36$, you can write:
$4 \times 90 = 360$; $36 \div 9 = 4$; $3.6 \div 4 = 0.9$; $0.4 \times 0.9 = 0.36$.

For each of these multiplications, write five derived number facts.

    a $6 \times 8 = 48$ **b** $7 \times 12 = 84$ **c** $3 \times 4.5 = 13.5$

1

> 28.56 − 24.7
> is 4.49

> 28.56 − 24.7
> is 3.86

Andrew states   Felicity states

Who is correct and what mistake has the other person made?

2 *Without* using a calculator, which of these three cards is the correct answer to this calculation, to 3 significant figures?

$$\frac{392 \times 6.32}{4.03^2}$$

242

189

153

Hint: Make your choice by estimating the answer.

3 This is a method for calculating the answer to a multiplication:

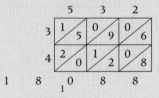

Hence 53.2 × 3.4 = 180.88

Can you find how this method works?
Use this method to work out 8.56 × 2.9.

4 Given that 165 × 46 = 7590, evaluate

a 1.65 × 0.46

b 75.9 ÷ 4.6

1 Work these out without using a calculator.

   a 421×12 **b** $4165 \div 5$

   c 178×46 **d** $4048 \div 16$

   e 623×28 **f** $29\,412 \div 12$

   g $15\,860 \div 52$ **h** 879×98

2 Expand the brackets and simplify, where possible.

   a $5(x + 4)$ **b** $p(p - 8)$

   c $3(a + 4) - 3(b + 2)$ **d** $6(h - 2) - 4(3 - h)$

   e $(x + 3)(x + 9)$ **f** $(y - 2)^2$

   g $(2a + 3)(a - 5)$ **h** $(p + q)(2p - 2q)$

3 Anna draws a rainbow.

The outer arc is a semicircle of radius 8 cm.
The inner arc is a semicircle of radius 6 cm.

Calculate the area of the rainbow.

Hint: The area of a semicircle is $A = \frac{1}{2}\pi r^2$. Find the area of the large semicircle and subtract the area of the small semicircle.

4 **a** Calculate
    i the mean
    ii the median
    iii the mode
    iv the range
    for this set of numbers.

    2, 2, 2, 2, 7, 8, 10, 11, 11, 12, 14

   b Explain why the mode may not be the best measure of the centre of the distribution in this case.

Example

The operations that have been applied to the equation

$$3x - 5 = 7$$

can be written as the function machine:

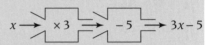

$$x \rightarrow \boxed{\times 3} \rightarrow \boxed{-5} \rightarrow 3x - 5$$

This equation is equal to 7, so beginning with 7, reverse the operations using the inverse function machine:

$$4 \leftarrow \boxed{\div 3} \leftarrow \boxed{+5} \leftarrow 7$$

$$x = 4$$

1 Use the function machine method to answer these.

  a $x + 8 = 15$ **b** $3x + 4 = 10$

  c $x^2 - 7 = 9$ **d** $\dfrac{x}{4} - 1 = 2$

  e $\dfrac{x + 7}{3} = 6$ **f** $3(2x - 5) = 21$

  g $\sqrt{5x - 4} = 6$ **h** $\dfrac{4(x + 3)^2}{8} = 50$

2 Solve these one-sided equations by algebraic methods.

  a $3x + 8 = 23$ **b** $5(y - 1) = 30$

  c $\dfrac{p}{3} - 4 = 1$ **d** $t^2 + 13 = 94$

  e $4(3c + 5) = 8$ **f** $\dfrac{7x - 6}{10} = 5$

  g $\dfrac{y^3 - 4}{15} = 4$ **h** $\sqrt{x} + 3 = 8$

3 The equation of a line is given by $y = 4x - 2$.

  a Work out the value of y when $x = 3$.

  b Work out the value of x when $y = -10$.

4 Given that the equation of a line is $y = 3x + c$, work out the value of c if the line passes through (2, 1).

1 Solve these equations.

a $2x + 9 = 3x + 4$ **b** $7p - 5 = 5p + 1$

c $4(2y + 1) = 6y + 8$ **d** $9a - 4 = 2(3a + 4)$

e $10 - 4x = 6x - 5$ **f** $5y + 7 = 9 + 7y$

g $4t - 3 = 12 - t$ **h** $4(3 - 4x) = 2(5 - 6x)$

2 Solve these equations.

a $\dfrac{x}{6} = 3$ **b** $\dfrac{a}{5} - 4 = 3$

c $\dfrac{8}{x} = 2$ **d** $\dfrac{4}{y} = 16$

e $\dfrac{3}{x} + 2 = 10$ **f** $4 - \dfrac{3}{e} = 1$

g $\dfrac{20}{x + 3} = 2$ **h** $\dfrac{10}{p + 3} = \dfrac{5}{p - 1}$

3 Write an equation for each of these statements and solve it to find the number that is being thought of.

a I think of a number, divide it by 5, add 3 and get 9.

b I think of a number, multiply it by 4 and add 3. I get the same answer as when I multiply my number by 2 and add 9.

c I think of a number, multiply it by 7 and subtract 4. I get the same answer as when I multiply my number by 2 and subtract it *from* 14.

d I think of a number, add 1, divide it *into* 15 and get 3.

4 Given that $y = \dfrac{10}{1 - x}$, find x when $y = -5$.

Example

Show the solution of $x \leqslant 3$ on a number line.

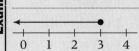

1 Solve these inequalities, representing your solution on a number line.

   a $2x < 9$

   b $3x + 5 \leqslant 23$

   c $5x - 4 \geqslant 11$

   d $-4x > 16$

   e $3(x - 8) < 6$

   f $2x + 4 \geqslant 5(x - 1)$

   g $10 \leqslant \dfrac{x}{3}$

   h $3(x - 1) > 7(x + 3)$

2 Simplify these expresssions.

   a $5x + 3y - 2x - y$

   b $4(3a - 1) - 3(a - 4)$

3 Solve these equations.

   a $5p + 8 = -2$

   b $15q + 2 = 3q - 2$

4 If $y = 6x + c$,

   a Work out the value of y when $x = -2$ and $c = 10$.

   b Work out the value of c when $y = 8$ and $x = 2$.

5 n is an integer.

   a Write the values of n which satisfy the inequality

   $$-1 \leqslant n < 4$$

   b Solve the inequality

   $$4x + 3 \geqslant 5$$

1 Write these sets of numbers in ascending order.

   a 85.6, 58.6, 6.85, 685, 5.68, 5.86, 0.586, 0.865

   b 0.124, 0.12, 0.142, 0.014, 1.42, 2.1, 0.41, 4.12

2 Evaluate these, given that $a = -2$.

   a $3a + 8$ **b** $5a - 1$

   c $10 - a$ **d** $\dfrac{a}{2}$

   e $a^2 - 3a + 2$ **f** $\dfrac{2a + 1}{3}$

   g $2a^3$ **h** $3a^2 - 2a$

3 Find the perimeter of these shapes.

a
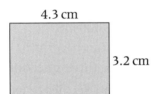
4.3 cm

3.2 cm

b
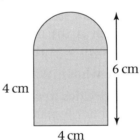
6 cm

4 cm

4 cm

4 PQRS is a trapezium.

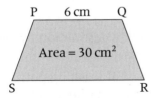

P   6 cm   Q

Area = 30 cm²

S   R

PQ is 6 cm and PQ : SR = 2 : 3.
The area of PQRS is 30 cm².
Find the distance between PQ and SR.

1 Suggest improvements to these questions for use in a questionnaire on music:

> **a** Do you agree that rock music is the best type of music?
>
> **b** How often do you buy CDs?
> ☐ Sometimes ☐ Often ☐ Never
>
> **c** What is you opinion of Radio 1? ☐ Fantastic ☐ Good
>
> **d** How many music concerts do you attend each year?
>
> **e** What is your full address?

2 Jaya decides to conduct a survey on school uniform. She writes these three questions

i Don't you think that all children should wear school uniform?

ii Do you not agree that it is not wrong to not allow children to attend school if they are not wearing uniform when it is school policy?

iii Which is the best colour for a school uniform?
☐ Green ☐ Blue

a Write what is wrong with each of these three questions.

b Jaya decides to carry out her survey and decides to ask five of her female friends. Write why this sample may be biased.

3 For each of these surveys, suggest reasons why the chosen sample may be biased.

a To find out the most popular holiday destination by asking people as they leave a travel agents.

b To find out the most popular brand of washing powder by asking people as they go in to a supermarket at 11 a.m. on a Tuesday.

c To find out how people travel to school by telephoning all the households on one page of the telephone directory on a Wednesday.

1 In her survey on school uniform, Jaya decides to concentrate on the difference of opinion between boys and girls at her school. Design a two-way table in which Jaya can collect her data.

2 Beris is conducting a survey to determine the preferred type of music in her school. She is interested in the effect that age has on the type of music that is listened to and decides to divide her results according to year group.
Design a two-way table in which Beris can collect her data.

3 This table gives information concerning the number of students in Years 7 to 9 that stay for a school dinner on 7 October.

	Year 7	Year 8	Year 9
Boys	77	65	58
Girls	63	60	57

a How many students in Years 7 to 9 had a school dinner on 7 October?

b Work out the percentage of students that were

  i girls   **ii** Year 7 boys.

c What percentage of the girls were in Year 8?

4 This is the data Priya collected on the amount of pocket money, in pounds, that she received over 15 weeks.

10   8   10   6   10   12   5   8   9   50   14   15   5   10   8

a Find, for this set of numbers
  i the mean   **ii** the mode
  iii the median   **iv** the range
  v the interquartile range

b Suggest a reason for why Priya received £50 one week. What effect did this have on each of your averages in part **a**?

1 A teacher has bought a new textbook for use in his lessons. He wants to find out what his students think of this change. He asks this question in a survey:

'What do you think of the new textbook?'

☐ Excellent ☐ Very Good ☐ Good

a Write what is wrong with this question.

This is another question in the survey:

'How much time do you normally spend on homework?'

☐ A lot ☐ A little

b i Write what is wrong with this question.
 ii Design a better question for the teacher to use.

2 Florence recorded her test results in the back of her exercise book.

Maths	English	French	Biology	Art
84%	80%	75%	77%	■%

She says, 'My mean mark in these five tests was 78%.'

Unfortunately, Florence noticed that there was an ink blot covering her mark in Art. Can you help her to work out her mark for Art?

3 There are 30 students in a class, 14 girls and 16 boys. On one particular night, the mean time spent on homework by the girls was 1.4 hours and the mean time spent on homework by the boys was 1.9 hours.

Work out the mean time spent on homework by all the students in this class. Give your answer in hours and minutes.

Hint: Work out the mean time as a mixed number not a decimal.

4 There are 18 girls and 14 boys in a class. In a French exam, the mean mark for the girls was x and the mean mark for the boys was y. Work out an expression for the mean mark of the class.

1 Round these numbers to the degree of accuracy given in brackets.

  a 32.4 (nearest whole)

  b 125.6 (nearest 10)

  c 0.865 (1 decimal place)

  d 392.56 (3 significant figures)

  e 0.0269 (2 significant figures)

  f 8506 (nearest 100)

  g 6.6752 (2 decimal places)

  h 59 940 (2 significant figures)

2 The equation of a line is given by $y = 8 - 3x$.

  a Find the value of y when $x = 4$.

  b Find the value of x when $y = 5$.

3 A circular mirror has a circumference of 157.1 cm. Find the area of the mirror.

Hint: Substitute 157.1 cm into the formula for the circumference of a circle and rearrange to find the radius.

4 This table gives information about the number of students in Years 11–13 that attended a school skiing trip.

	Year 11	Year 12	Year 13
Male	8	14	17
Female	4	12	20

  a How many students took part in the skiing trip?

  b What percentage of these students were male?

  c Of the females that took part in the trip, what percentage were in Year 13?

Do not use a calculator for these questions.

1 Copy and complete these fraction equivalents.

a i $\dfrac{5}{7} = \dfrac{}{14}$ **ii** $\dfrac{}{9} = \dfrac{6}{27}$

iii $\dfrac{3}{} = \dfrac{15}{20}$ **iv** $\dfrac{7}{8} = \dfrac{63}{}$

b Find a fraction equivalent to $\frac{5}{12}$ whose denominator is a square number and whose numerator is *not* divisible by 2.

2 Write each list of fractions in ascending order.

a $\dfrac{1}{4}, \dfrac{2}{5}, \dfrac{11}{20}, \dfrac{1}{2}, \dfrac{3}{10}$ **b** $\dfrac{1}{5}, \dfrac{11}{15}, \dfrac{4}{45}, \dfrac{2}{3}, \dfrac{5}{9}$

c $\dfrac{5}{12}, \dfrac{4}{9}, \dfrac{1}{6}, \dfrac{3}{4}, \dfrac{1}{3}$ **d** $\dfrac{7}{10}, \dfrac{2}{15}, \dfrac{5}{6}, \dfrac{2}{3}, \dfrac{4}{5}$

3 Work out these calculations.

a $\frac{2}{7} + \frac{3}{7}$ **b** $\frac{5}{9} - \frac{2}{9}$

c $\frac{1}{2} + \frac{1}{3}$ **d** $\frac{7}{8} - \frac{3}{4}$

e $\frac{3}{5} + \frac{1}{4}$ **f** $1\frac{1}{3} - \frac{5}{12}$

g $2\frac{3}{4} + 1\frac{5}{6}$ **h** $4\frac{4}{9} - 2\frac{1}{6}$

4 Davina spent $\frac{1}{8}$ of her pocket money on chocolate, $\frac{2}{5}$ on a day out with friends and used the rest to buy her friend Karen a birthday present. What fraction of her money did Davina spend on Karen?

Do not use a calculator for these questions.

1 **a** True or False: Dividing by 8 is the same as multiplying by $\frac{1}{8}$. Explain your answer.

b Rewrite each of these divisions as a multiplication involving fractions. Work out the answers as mixed numbers.

i $9 \div 4$ **ii** $15 \div 7$

iii $12 \div 5$ **iv** $24 \div 9$

2 Work out these calculations.

a $\frac{3}{4} \times \frac{1}{8}$ **b** $\frac{3}{5} \times \frac{5}{9}$

c $\frac{5}{8} \div \frac{2}{3}$ **d** $\frac{5}{12} \div \frac{1}{6}$

e $\frac{7}{15} \times 12$ **f** $10 \div \frac{2}{3}$

g $2\frac{2}{5} \times 3\frac{1}{3}$ **h** $4\frac{2}{7} \div 3\frac{1}{8}$

3 Pair these cards together if they show equivalent numbers. Which is the odd card out? You should be able to do this question mentally.

$\frac{3}{5}$	0.3	$\frac{3}{8}$	0.35	0.6

$\frac{3}{4}$	0.375	$\frac{3}{10}$	0.75

4 Write each of these fractions as hundredths. Then convert each fraction to **i** a decimal and **ii** a percentage.

a $\frac{1}{4}$ **b** $\frac{4}{5}$ **c** $\frac{9}{20}$ **d** $\frac{8}{25}$

5 Use short division to convert each of these fractions to decimals.

a $\frac{3}{8}$ **b** $\frac{4}{9}$ **c** $\frac{5}{16}$ **d** $\frac{7}{12}$

N3 HW4 · Unit review

Do not use a calculator for these questions.

1 Chris said,

> I have two-thirds of a bottle of pop.

Julie said,

> I have six-eighths of a bottle of pop and my bottle of pop is exactly the same size as yours.

Who has got the most pop, Chris or Julie?
Explain your answer.

2 a Convert 55% to

  i a decimal number   **ii** a fraction in its simplest terms.

  b Convert the recurring decimal $0.\dot{5}$ to a fraction.

3 a Find the lowest common multiple of 20, 25 and 50.

  b List these numbers in descending order.

  $\frac{3}{25}$, 50%, 0.6, 3%, $\frac{47}{50}$, 1, $\frac{9}{20}$, 25%

4 Two lengths of piping are fastened together.

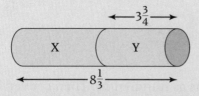

The combined length of the piping is $8\frac{1}{3}$ inches.
The length of piping marked Y is $3\frac{3}{4}$ inches.

Work out the length of the piping marked X.

1 Copy and complete this equivalence table.

Fraction	Decimal	Percentage
$\frac{3}{4}$		
	0.4	
		35%
	0.05	
$\frac{5}{8}$		

2 Solve these inequalities.

a $x - 5 > 7$ **b** $3 + x \leqslant 9$

c $2 + n \geqslant -3$ **d** $2(x + 1) \leqslant x - 4$

e $\dfrac{x}{3} \geqslant 4$ **f** $\dfrac{2x}{7} < 4$

g $10 < 3x < x + 7$ **h** $x^2 \leqslant 25$

3 Calculate
   i the area and
   ii the perimeter of a circle of
     diameter 14 m.

14 m

4 Kirsty carries out a survey to find out the popularity of canoeing at her school. She asks all the students in Year 11.

a Explain why this sample could be biased.

b Suggest how Kirsty could choose an unbiased sample.

Kirsty decides to find out if other students would like to go on a canoeing trip to Alaska. She is interested in any difference of opinion between boys and girls.

c Design a two-way table for Kirsty to collect her data.

1 Work out the missing angles, giving reasons for your answers.

a

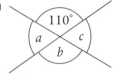

b

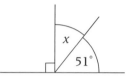

c

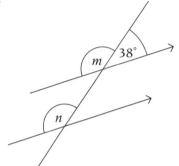

d

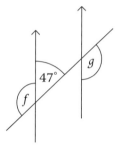

2 Work out the missing angles, giving reasons for your answers.

a

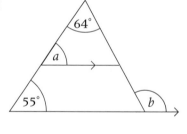

b

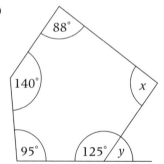

3 Calculate the interior angle sum of a hexagon, using diagrams to show your working.

Hint: Divide the hexagon into triangles first.

1 Work out the missing angles, giving reasons for your answers.

a

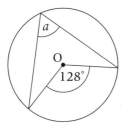

b

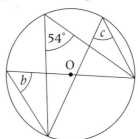

c

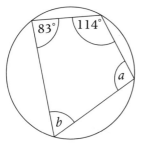

d

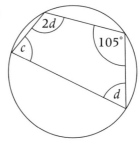

2 Work out the missing angles, giving reasons for your answers.

a

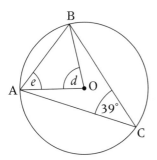

b

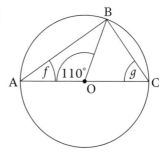

c

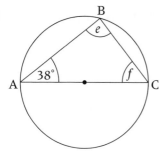

d

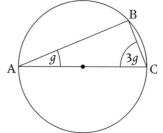

1 Work out the missing angles, giving reasons for your answers.

a

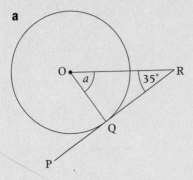

b

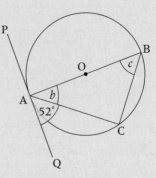

2 A, B and C are points on the circumference of a circle, centre O.

  a Find angle AOC.

  b Give a reason for your answer.

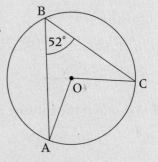

3 a The exterior angle of a regular polygon is 30°. Write the interior angle of this polygon.

  b The exterior angle of a regular polygon is 40°. Work out the number of sides of this polygon.

  Name this polygon.

1 a Convert these recurring decimals to fractions.

    i 0.777 777 ...   **ii** 0.494 949 49 ...

    iii $0.7\dot{3}$ **iv** 0.345 345 345 ...

Hint: Write each fraction in its simplest form.

b Convert these fractions to recurring decimals.

    i $\frac{4}{9}$ **ii** $\frac{5}{6}$ **iii** $\frac{1}{7}$

Hint: Write $\frac{4}{9}$ as the division $9\overline{)4.0000}$

2 Solve these equations.

    a $\dfrac{x}{5} = 7$ **b** $\dfrac{12}{x} = 3$

    c $\dfrac{6}{x} - 1 = 2$ **d** $\dfrac{x}{9} + 4 = 8$

3 Find the area of these shapes.

a

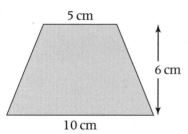

5 cm · 6 cm · 10 cm

b

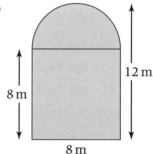

12 m · 8 m · 8 m

4 In Class A the mean average score in a French exam was 72%. In Class B the mean average score in a French exam was 84%. There are 30 students in Class A and 26 students in Class B.

Calculate the mean average score for the two classes combined. Give your answer to 3 significant figures.

1 Write the next two terms in each of these number patterns.

   a 4, 8, 12, 16, 20, ...   **b** 3, 7, 11, 15, 19, ...

   c 100, 94, 88, 82, 76, ...   **d** 9, 16, 25, 36, 49, ...

   e 2, 4, 8, 16, 32, ...   **f** 1000, 500, 250, 125, 62.5, ...

   g 1, 1, 2, 3, 5, 8, ...   **h** 1, 2, 6, 24, 120, ...

2 Look at this number pattern.

$$9^2 = 81$$
$$99^2 = 9801$$
$$999^2 = 998\ 001$$
$$9999^2 = 99\ 980\ 001$$

   a Write the next two lines in this pattern. Use your calculator to check your answers.

   b By relating the number of digits that are 9 in the question to the number of digits that are 9 in the answer, work out $99\ 999\ 999^2$.

3 Look at this number pattern.

$$1^2 = 1$$
$$2^2 = 1 + 3$$
$$3^2 = 1 + 3 + 5$$
$$4^2 = 1 + 3 + 5 + 7$$

   a Write the next three lines in this pattern.

   b Work out the value of k if

$$50^2 = 1 + 3 + 5 + \cdots + k$$

4 Generate the first five terms of each of these sequences.

   a $T_n = 4n - 3$ **b** $T_n = 50 - 3n$ **c** $T_n = n^2 + 5$

   d $T_n = (n + 1)(n + 3)$ **e** $T_n = \dfrac{n}{n + 1}$ **f** $T_n = \dfrac{1}{n^2}$

1 Find the *n*th term of these linear sequences.

   a 7, 9, 11, 13, 15, ...   **b** 3, 6, 9, 12, 15, ...

   c 3, 8, 13, 18, 23, ...   **d** −8, −4, 0, 4, 8, ...

   e 17, 16, 15, 14, 13, ...   **f** 20, 17, 14, 11, 8, ...

   g $1\frac{1}{2}$, 2, $2\frac{1}{2}$, 3, $3\frac{1}{2}$, ...   **h** $4\frac{3}{4}$, $4\frac{1}{2}$, $4\frac{1}{4}$, 4, $3\frac{3}{4}$, ...

2 Choose one of these *n*th terms for each of these sequences.

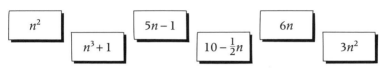

       n^2 $5n-1$ $6n$

           n^3+1 $10-\frac{1}{2}n$ $3n^2$

   a $9\frac{1}{2}$, 9, $8\frac{1}{2}$, 8, $7\frac{1}{2}$, ...   **b** 2, 9, 28, 65, 126, ...

   c 6, 12, 18, 24, 36, ...   **d** 1, 4, 9, 16, 25, ...

   e 4, 9, 14, 19, 24, ...   **f** 3, 12, 27, 48, 75, ...

3 **a** In Class 1, the tables are arranged with the short sides
     pushed together. Work out a formula connecting the
     number of tables (*t*) with the number of students (*S*)
     seated around them.

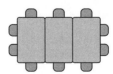

   b In Class 2, the tables are arranged with the long sides
     pushed together. Work out a formula connecting the
     number of tables (*t*) with the number of students (*S*)
     seated around them.

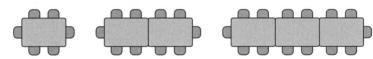

   c Explain **why** each formula works.

1 Find the nth term for each of these quadratic sequences.

  a 2, 8, 18, 32, 50, ...   **b** 4, 7, 12, 19, 28, ...

  c 6, 14, 24, 36, 50, ...   **d** 3, 13, 29, 51, 79, ...

2 Nathan has drawn a pattern. The graph shows the number of dots, d, used in each pattern, n.

  a Write a formula for d in terms of n.

  b How many dots did Nathan use in pattern 8?

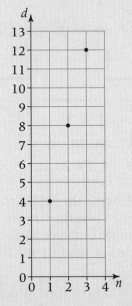

3 The first 5 terms of an arithmetic sequence are

  5,   9,   13,   17,   21

  Find an expression, in terms of n, for the nth term of the sequence.

4 This sequence does not start with term one. Find the nth term of the sequence.

  15th term = 117
  16th term = 125
  17th term = 133

1 Use short division to convert each of these fractions to decimals.

a $\frac{5}{8}$ **b** $\frac{1}{6}$ **c** $\frac{1}{16}$ **d** $\frac{8}{9}$ **e** $\frac{5}{12}$

2 Copy the diagram and draw lines to pair each sequence with its nth term. The first one has been done for you.

Sequence		nth term
4, 3, 2, 1, 0, …		$4n^2$
4, 7, 10, 13, 16, …		$4n$
4, 9, 16, 25, 36, …		$5-n$
4, 8, 12, 16, 20, …		$3n+1$
4, 16, 36, 64, 100, …		$(n+1)^2$

3 Work out the missing angles, giving reasons for your answers.

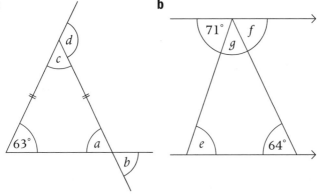

a

b

4 Lydia recorded a list of the five times in which she ran a 100 m sprint in her last athletics season. Unfortunately, one of her times went missing. Find Lydia's missing time from the information given.

Times (in s) for the 100 m sprint: | 14.3 | 14.1 | 13.9 | 14.2 |
Mean time: 14.2 seconds

1 The table gives the mean temperature (°C) in England and Wales for each month in 2004 and the number of ice cream cornets sold at Mr Frosty's icecream parlour in Lincolnshire.

Month	Jan	Feb	Mar	Apr	May	Jun
Temperature	4.9	5.0	6.2	9.1	11.7	15.0
Cornets	15	18	20	52	115	164

Month	Jul	Aug	Sep	Oct	Nov	Dec
Temperature	15.4	17.1	14.4	10.3	7.5	5.3
Cornets	200	188	124	80	28	16

a Represent these data on a scatter diagram.
b Describe the correlation shown in terms of the data.

2 The table gives the marks of 10 students in the two papers of a French exam.

Paper 1	75	64	50	80	45	58	72	63	50	74
Paper 2	81	70	52	89	42	62	72	73	48	79

a Represent these data on a scatter diagram.
b Describe the correlation shown in terms of the data.
c Draw a line of best fit and use this to estimate a mark in paper 1 for a student who was absent on the day of the exam but who scored a mark of 55 on paper 2.

3 Danielle chose a random sample of 12 people and recorded their shoe size and their height in centimetres.

Shoe size	5	6	9	12	5	5
Height	155	170	180	188	156	160

Shoe size	8	6	11	8	10	4
Height	168	165	183	172	180	152

a Represent these data on a scatter diagram.
b Describe the correlation shown in terms of the data.
c Draw a line of best fit and use this to estimate the height of a person who has a shoe size of 7.

1 These are the Physics test results of a group of 25 students.

77	54	85	21	62	95	87	50	93
49	65	51	84	75	77	48	69	64
72	77	70	68	65	91	89

  a Represent these data on an ordered stem-and-leaf diagram.
  b Describe the trend.
  c Find the mode and the median of these data.

2 These are the heights, in cm, of a group of 30 sixth form girls.

160	154	155	171	165	158	152	170	163	164
160	162	157	159	163	169	161	166	174	156
167	150	168	154	158	171	164	158	158	162

  a Copy and complete this stem-and-leaf diagram.

15	4
15	5
16	0
16	5
17	1
17	

The first five numbers have been done for you

  b Draw an ordered stem-and-leaf diagram.
  c Find the range and the interquartile range of these data.

3 These are the mean temperatures (in °C) for March recorded in England and Wales during the last two decades of the 20th century.

1980 to
1989:   4.3   7.4   5.6   6.0   4.3   4.3   4.6   3.8   5.9   7.0

1990 to
1999:   7.9   7.4   6.9   6.1   7.1   5.0   4.0   7.8   7.5   6.9

  a Represent these data on a back-to-back stem-and-leaf diagram.
  b Compare the two sets of data. You may like to mention the median, mode, range or interquartile range.

1 Look at these scatter graphs.

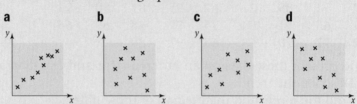

a b c d

Which scatter graph could represent the relationship between
   i   Height and shoe size?
   ii Temperature and ice-cream sales?
   iii IQ and house number?
   iv Price and mileage of Ford cars?

2 The table shows the infant mortality rate (per 100 live births) and the life expectancy, in years, (at birth) of 10 countries.

Country	Infant mortality rate	Life expectancy
Australia	4.7	80.4
Denmark	4.6	77.6
Germany	4.2	78.7
Ireland	5.4	77.6
Japan	3.3	81.2
Norway	3.7	79.4
Poland	8.5	74.4
Sweden	2.8	80.4
UK	5.2	78.4
USA	6.5	77.7

a Plot this information on a scatter graph.
b Describe the relationship between a country's infant mortality rate and its life expectancy.

3 Here are the times, in minutes, taken to complete some crossword puzzles.

12   15   21   10   32   9   18   24   20   8
26   22   19   11   28   23   23   14   25   27

a Represent these data on a box plot.
b Find the median and interquartile range of these data and comment on any trend.

1 Work out these fraction additions and subtractions.

a $\frac{1}{2} + \frac{1}{4}$ **b** $\frac{1}{5} + \frac{3}{10}$

c $\frac{5}{6} - \frac{1}{4}$ **d** $\frac{3}{5} + \frac{1}{3}$

e $\frac{5}{9} - \frac{1}{12}$ **f** $\frac{4}{7} + \frac{3}{8}$

g $\frac{4}{5} - \frac{2}{3}$ **h** $\frac{7}{10} + \frac{7}{15}$

2 Solve these equations, with unknowns on both sides.

a $x + 9 = 3x - 1$

b $5q - 4 = 2q - 10$

c $3(a + 4) = 5(2a + 1)$

d $10 - 3y = 4(1 - y)$

3 Moggie says:

'I think of a number, add 2, divide it *into* 30 and get 5.'

Write an equation for this statement and solve it to find the missing number.

4 Calculate the sum of interior angles of a nonagon, using diagrams to show your working.

5 Lynda recorded the Mathematics test results of 30 students.

```
38   89   52   42   40   35   62   51   54   53
77   71   84   55   65   62   80   74   70   43
31   48   56   66   64   72   70   79   87   82
```

Represent these data on a stem-and-leaf diagram and find the median mark in this Mathematics test.

1 Copy and complete these tables to draw these straight-line graphs.
Draw each line on a separate set of axes.
Label each set of axes from −8 to +8 in both the x and y directions.

a $y = 2x + 3$

x	−2	0	2
y	−1		

b $y = 4 - x$

x	−3	0	3
y		4	

Hint: −− makes a +

c $2y = x - 4$

x	1	2	3
y			$-\frac{1}{2}$

d $y = 3$

x	−1	0	1
y			

2 Which of these lines passes through the point (3, 4)?

Line	✓ or ✗
$y = x + 1$	
$y = 3x - 5$	
$2y = x - 1$	
$3y = 2x - 2$	

3 a On a set of axes labelled from 0 to 8 in both the x and y directions, plot the graphs of $x = 2$ and $y = 6$. Where do these lines intersect?

b *Without* drawing the graphs, write where these pairs of lines intersect.

  i $x = 4$ and $y = 3$
  ii $y = -3$ and $x = 5$
  iii $x = \frac{3}{4}$ and $y = \frac{5}{8}$

Example

Find the gradient and y-intercept of $y = 6x - 2$.

Gradient = 6, y-intercept = -2

1 a Copy and complete this table to draw the straight-line graph $y = 2x - 1$.
Use axes numbered from -6 to $+6$ in both the x and y directions.

x	-2	0	2
y			

b Write

i the coordinate of the y-intercept

ii the gradient of the line.

c Write the connection between the characteristics identified in part **b** and the equation of the line $y = 2x - 1$.

2 Write

i the coordinate of the y-intercept

ii the gradient

of each of these straight-line graphs.

a $y = 3x + 4$ **b** $y = 5x - 1$

c $y = 6 - 3x$ **d** $x + y = 2$ Hint: $y = ?$

e $2y = 8 - x$ **f** $4x - 2y = 3$

3 Write the equation of a line that is parallel to

a $y = 2x + 5$
b $y = 4 - 3x$
c $x + 2y = 1$

4 Find the equation of a line that is parallel to $y = 4 - x$ and cuts the y-axis at $(0, 1)$.

1 Write the equations of these three lines.

  a Line A has a gradient of 3 and crosses the y-axis at $(0, 2)$.

  b Line B has a gradient of 4 and passes through $(1, -1)$.

  c Line C passes through $(0, 7)$ and $(3, -5)$.

2 Draw the graph of $y = 3x - 10$.

  Hint: Choose three values of x and construct a table.
  Find the corresponding values of y. Think carefully about the axes.

3 Here are the equations of five straight lines. Write the equations of two lines that are parallel to each line.

  A $y = 3x + 5$ **C** $3y + x = 5$ **E** $3y - x = 5$

  B $y = 5 - 3x$ **D** $3y = x - 5$

4 A straight line has equation $y = \frac{1}{4}x - 2$.

  The point P lies on the straight line. P has a y-coordinate of -1.

  a Find the x-coordinate of P.

  b Write the equation of a straight line that is parallel to $y = \frac{1}{4}x - 2$.

5 Where does the line $3y = 7x + 5$ cross

  a the y-axis
  b the x-axis?

1 Work out each of these questions using written methods.

   a $6.98 + 4.5$ **b** $12.05 - 3.27$

   c $3.52 + 0.596$ **d** $23.5 - 19.41$

2 Write

   i the coordinate of the y-intercept
   ii the gradient

   of each of these straight-line graphs.

   a $y = x + 3$ **b** $y = 5 + 4x$ **c** $y = 8 - x$

   d $y - 3x = 2$ **e** $x + 2y = 3$ **f** $9x - 3y = 5$

3 Work out the surface area of these prisms.

   a 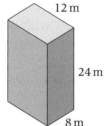   **b**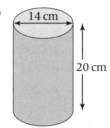

   Hint: The circumference of one end is the length of the rectangle
   that is folded to form the curved surface.

4 Evan collected data on the second-hand price, given to 2
   significant figures, of Ford Focus 1.8 TDCi cars.

Years old	1	1	2	2	3	3	4	5	5	6
Price (in thousands)	8.6	9.6	8	8.5	7.4	6	6.8	5	5.4	4.5

   a Plot these data on a scatter diagram and describe any
      correlation.

   b Draw a line of best fit and use this to estimate the price of
      a second-hand Ford Focus 1.8 TDCi which is 3.5 years old.

   c Comment on the reliability of your line of best fit and
      estimate.

1 A tin contains 2 blue pencils, 4 red pencils and 6 green pencils. A pencil is chosen at random. What is the probability that the pencil is

   a red   **b** blue   **c** not green   **d** yellow?

2 A credit card company offers a choice of designs for their cards. The table shows the picture on each card and the related probability.

Shoes	Diamonds	Tiger	Sports Car
0.21	x	0.28	0.19

Find the value of x.

3 A bag contains 15 cards; 6 are printed with a picture of a star, 4 with the Moon, 3 with planet Earth and 2 with a comet. A card is selected at random. What is the probability that the card chosen is

   a a star   **b** not a comet

   c the Milky Way   **d** the Moon or the Earth

   e a star, the Moon or a comet

   f a star, the Moon, the Earth or a comet?

4 Twenty balls, numbered 1 to 20, are placed in a box. A ball is selected at random.

   a What is the probability that the number on the chosen ball is

     i   a 3   **ii**   divisible by 5

     iii   a multiple of 4   **iv**   greater than 10

     v   a prime number   **vi**   not an odd number

     vii a 4 or less   **viii** a 7 or an even number?

   b What can you say about the pair of events 'the number on the ball is a 7 or the number on the ball is even'?

1 A group of 40 students listened to three pieces of classical music and their preferred piece was noted. The table shows this information.

	Bach	Chopin	Debussy	Total
Male		6		18
Female	4			
Total		14	12	40

a Copy and complete the table.

b One student is chosen at random. Find the probability that the student

  i is female   **ii** prefers Debussy
  iii is male and prefers Chopin.

c These 40 students were a random sample selected from a year group of 140 students. How many of the year group would you expect

  i to prefer Bach   **ii** to be male?

2 A biased dice is rolled 150 times. The table shows the outcomes.

Score	1	2	3	4	5	6
Frequency	50	24	22	20	26	8

a If the dice is rolled once more, estimate the probability that it will land on a
  i 1   **ii** 6   **iii** 2 or 5.

b The dice is to be rolled 300 times. How many times would you expect the dice to land on a
  i 2   **ii** 3 or 4?

3 Suzanne decides to roll a biased dice 500 times. The probability that the dice will land on a six is 0.45.

Work out an estimate for the number of times that the dice will land on a six.

1 Freya carries out a statistical experiment. She throws a dice 120 times. She scores a three 40 times.
Is the dice fair? Explain your answer.

2 Georgia plays a game on the computer. The probability that she completes each level of the game is shown in the table.

Level 1	Level 2	Level 3	Level 4
$8x$	$4x$	$2x$	x

  a Comment on the difficulty of each level in comparison to the previous one.
  b Work out the probability of completing each level.

3 The probability that Ethan walks to school is $\frac{5}{8}$ and the probability that he takes the bus is $\frac{1}{4}$. At all other times, Ethan's parents drive him to school in their car. Find the probability that
  a Ethan's parents drive him to school
  b Ethan does not walk to school
  c Ethan takes either the bus or walks to school.

4 A group of 50 Maths teachers are asked to choose their favourite section of the GCSE Maths syllabus. This table shows their choices.

	Number	Algebra	Shape	Data	Total
Male	3			6	
Female		10	3	4	
Total	10		12		50

  a Copy and complete the table.
  b A Maths teacher is chosen at random. Find the probability that

   i the teacher prefers Algebra
   ii the teacher is a male who prefers Data.

  c If these teachers are a sample of 500 teachers attending a Mathematics convention, how many of the larger group would you expect to prefer Shape?

1 Given that $6.95 \times 540 = 3753$, write the answer to

   a 69.5×54 **b** $6.95 \times 54\,000$

   c $37.53 \div 54$ **d** 0.695×5.4

   e $3.753 \div 0.54$

2 Given that the equation of a line is $y = c - 2x$, work out the value of c if the line passes through $(3, -3)$.

3 Find the missing angles, giving reasons for your answers.

a

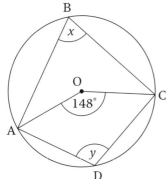

b

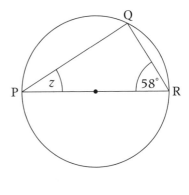

4 Daniel and Hannah are writing out Christmas cards. They record the time (in seconds) that it takes to complete each card and envelope and compile the information shown in the table.

	Daniel	Hannah
Median	55 s	48 s
Lower quartile	44 s	40 s
Upper quartile	58 s	60 s
Minimum	35 s	33 s
Maximum	61 s	68 s

   a Draw box plots for each set of information on the same axes.

   b Comment on and compare the performances of Daniel and Hannah.

N4 HW2 Direct proportion

1 Kent surveys his team of 30 engineers. He notes that 25 of his team really enjoy their job and that 4 of his team have worked in other professions before becoming engineers.

   a What proportion of the engineers really enjoy their job?

   b What proportion of Kent's team have only ever worked as engineers?

2 Ash has a paper round. She delivers $\frac{1}{4}$ of her newspapers to Scarlet Street and $\frac{2}{5}$ to Ruby Road.
Ash delivers the rest of her newspapers to the houses in the street where she lives.

   a What proportion of the newspapers does Ash deliver to the houses in the street where she lives?

   b If Ash delivers 160 newspapers in total, how many does she deliver to
     i Ruby Road
     ii her own street?

3 A 7 m length of fabric costs £196. Find the cost of 4 m of the same fabric. Show your working.

Hint: Use the unitary method to find the price of one metre of fabric.

4 Fabric conditioner is sold in two differently sized bottles.
A standard bottle has a capacity of 750 ml and costs £1.26.
A large bottle has a capacity of 1.25 litres and costs £1.99.

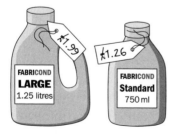

Which bottle is the best value for money? Show your working.

1 Johanna is taking a round the world trip and decides to convert £115 into the local currency of each place that she intends to visit. Copy and complete the table.

Country	Exchange rate		Amount of local currency
Australia	1 GBP =	Australian dollars	AU$269.10
Bolivia	1 GBP = 14.2	Bolivian bolivianos	
France	1 GBP =	Euros	€169.05
Iceland	1 GBP = 111.43	Icelandic krona	
Maldives	1 GBP = 22.62	Maldives ruiyan	
Thailand	1 GBP = 46.78	Thailand Baht	

2 Laura decides to go on holiday to New Zealand. She converts £500 into New Zealand dollars at an exchange rate of

> 1 GBP = 2.6 New Zealand dollars

a Work out how many New Zealand dollars she receives.

b Unfortunately, Laura has to cancel her holiday and converts the full amount back to pounds sterling at an exchange rate of

> 1 GBP = 2.56 New Zealand dollars

Without actually calculating the amount, will Laura make a profit or loss from this final exchange?

3 A car travels 51 miles in 1 hour and 25 minutes. Find the average speed of the car in miles per hour.

4 A block of aluminium is in the shape of a cuboid of dimensions 3.4 m by 0.5 m by 2.1 m. Aluminium has a density of 2.7 g/cm^3. Work out the mass of the block of aluminium in kilograms.

Hint: Work out the volume in m^3 and then convert to cm^3.

1 Harriette travelled to America and bought a pair of designer jeans.

She paid \$132.75.

Back home in England, Harriette sees an identical pair of jeans for £95.

If the exchange rate is £1 = \$1.77 did Harriette make a saving and if so, how much? Give your answer in pounds.

2 A solid iron bar is in the shape of a cuboid of width 2 cm, height 12 cm and length 30 cm.

Iron has a density of 7.87 g/cm^3.

Work out the mass of the iron bar in kg to 3 significant figures.

3 Simon is shopping for a new notebook computer. He decides on one particular brand and model and discovers that it is on offer in three different shops.

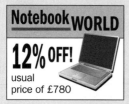  

Notebook WORLD
12% OFF!
usual price of £780

Asteroid
$\frac{1}{5}$ **OFF**
usual price of £850

Balti's
£580
plus VAT at 17.5%

Find the cost of the notebook computer in all three shops and decide on the best offer for Simon.

4 Jake has £3500 in a bank account that earns 4.5% compound interest.

Rowan has £3500 in a bank account that earns 4.7% simple interest.

After 3 years, who will have the most money in their account assuming that no withdrawals have been made?

1 Write these fractions, decimals and percentages in ascending order.

  a 0.45, 23%, $\frac{41}{50}$, 1, 0.12, 95%, $\frac{1}{20}$, $\frac{4}{5}$

  b $\frac{1}{2}$, 0.24, $\frac{39}{50}$, 90%, 0.3, 2%, $\frac{17}{100}$, 65%

2 **a** **i** Find the first five terms of the sequence given by

  $$u_n = 19 - 2n$$

  ii For which value of n is $u_n = -11$?

  b Find the general term of the sequence given by

  $$3, 7, 11, 15, 19, \dots$$

3 **a** Find the size of an exterior angle of an octagon, showing your working.

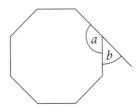

  b Find the size of an interior angle of an octagon, showing your working.

4 Suggest improvements to these questions for use in a questionnaire.

> **a** How many hours of television do you watch each day?
>
> **b** Don't you think it is wrong to let a child watch more than 4 hours television a day?

1 Copy this diagram.

a Reflect the triangle A
   in the line $x = 1$.

   Label the image B.

b Reflect the triangle A
   in the line $y = 1$.

   Label the image C.

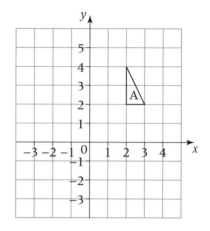

2 Using square grid paper, draw a set of axes.
Label the x-axis from 0 to 7.
Label the y-axis from −3 to 7.

Draw a rectangle with vertices (3, 1), (3, 2), (6, 2), (6, 1).
Label the rectangle R.

a Reflect the rectangle R in the line $y = 0$. Label the image P.

b Reflect the rectangle R in the line $y = x$. Label the image Q.

3 Using square grid paper, draw a set of axes.
Label the x-axis from −3 to 6.
Label the y-axis from −5 to 5.

Draw a shape with vertices (2, 1), (2, 4), (3, 4), (3, 2), (4, 2),
(4, 1).
Label the shape A.

a Rotate shape A 90° clockwise about (1, 0).
   Label the image B.

b Rotate shape B 90° clockwise about (1, 0).
   Label the image C.

c Describe a single transformation that maps shape A
   onto shape C.

S3 HW3 Translations and describing transformations

1 Write vectors to describe these translations.

a Triangle A to triangle B

b Triangle A to triangle C

c Triangle B to triangle D

d Triangle C to triangle D

e Triangle D to triangle E

f Triangle E to triangle C

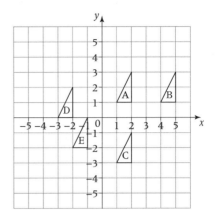

2 Using square grid paper, draw a set of axes.

Label the x-axis from −5 to 8.

Label the y-axis from 0 to 5.

Draw a parallelogram with vertices (1, 1), (2, 2), (5, 2), (4, 1).

Label the parallelogram P.

a Translate the parallelogram P through $\begin{pmatrix} -5 \\ 1 \end{pmatrix}$. Label the image Q.

b Translate the parallelogram P through $\begin{pmatrix} 2 \\ 2 \end{pmatrix}$. Label the image R.

3 Describe fully the transformation that maps

a Triangle A to triangle B

b Triangle C to triangle B

c Triangle B to triangle D

d Triangle D to triangle A

e Triangle E to triangle A

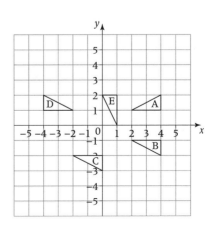

1 Using square grid paper, draw a set of axes.

Label the x-axis from 0 to 6.

Label the y-axis from −3 to 4.

Draw a rectangle with vertices (2, 1), (2, 3), (3, 3), (3, 1).

Label the rectangle A.

T is a translation with column vector $\begin{pmatrix} 2 \\ -3 \end{pmatrix}$.

a Perform T on rectangle A. Label the image B.

b Find the translation T^{-1} that maps shape B onto A.

c What do you notice about these two column vectors?

2 a Reflect triangle A in the line $y = 0$.
Label the image B.

b Reflect triangle B in the line $y = x$.
Label the image C.

c Describe fully the single transformation that maps triangle A onto triangle C

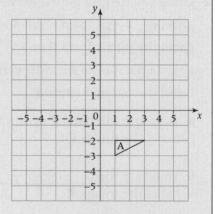

3 a Reflect shape L in the line $x = 0$. Label the image M.

b Reflect shape M in the line $x = 3$. Label the image N.

c Describe fully the single transformation that maps shape L onto shape N.

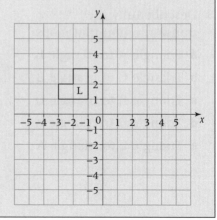

1 If a 1.5 m length of copper piping has a mass of 9.9 kg, work out the mass of these lengths of the same copper piping.

   a 0.5 m   **b** 3 m

   c 5 m   **d** 12 m

2 Find the equation of the straight line joining these pairs of points.

   a $(0, 1)$ and $(3, 7)$

   b $(-1, -9)$ and $(5, 9)$

   c $(2, 1)$ and $(4, -1)$

3 **a** Reflect shape F in the line $y = 0$. Label the image G.

   b Reflect shape G in the line $y = 3$. Label the image H.

   c Describe fully the single transformation that maps shape F onto shape H.

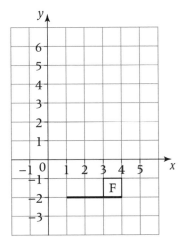

4 A dice is rolled.

   a Explain why the events 'obtain an odd number' and 'obtain a 2' are mutually exclusive.

   b Find the probability of obtaining

     i   an odd number
     ii a 2
     iii an odd number or a 2
     iv neither an odd number nor a 2.

Example

Work out the value of
$3^3 \times 3^2 \div 3$

$3^3 \times 3^2 \div 3 = 3^{(3+2)} \div 3 = 3^{(5-1)} = 3^4 = 81$

1 Write

   a 8 as a power of 2   **b** 64 as a power of 4

   c 81 as a power of 3   **d** 100 000 as a power of 10

   e 1296 as a power of 6   **f** $\frac{1}{4}$ as a power of $\frac{1}{2}$

2 Work out these, giving your answer in index form.

   a $2^5 \times 2^3$ **b** $4^3 \times 4^6$

   c $8^6 \times 8$ **d** $x^2 \times x^7$

   e $5^9 \div 5^4$ **f** $10^7 \div 10^6$

   g $a^8 \div a^8$ **h** $9^4 \div 9^7$

   i $3^2 \times 3^2 \times 3^2$ **j** $y^2 \times y^3 \times y^4$

   k $6^5 \div 6^2 \times 6^4$ **l** $7^6 \times 7^4 \div 7^3$

3 Work out the value of each of these.

   a 10^0 **b** $(2^4)^2$

   c $(\sqrt{5})^2$ **d** 2005^1

4 Simplify these expressions, giving your answer in index form.

   a $\dfrac{8^6}{8^3}$ **b** $\dfrac{4^2 \times 4^5}{4^3}$

   c $\dfrac{3^9}{3^2 \times 3^4}$ **d** $\dfrac{x^4 \times x^5}{x^2 \times x}$

   e $7^3 \times \dfrac{7^7}{7^4 \times 7^5}$ **f** $\dfrac{(a^2)^3}{a^2 \times a^3}$

1 Simplify these expressions, giving your answer in index form.

   a $(3^2)^4$ **b** $(5^3)^2$ **c** $(9^{\frac{1}{2}})^4$ **d** $(8^4)^{\frac{1}{3}}$

2 Pair these cards together if they show equivalent numbers. Which is the odd card out?

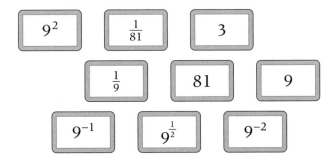

3 Evaluate each of these.

   a 12^0 **b** $25^{\frac{1}{2}}$ **c** 3^{-1}

   d $27^{\frac{1}{3}}$ **e** 2^{-3} **f** 9^{-2}

   g $36^{-\frac{1}{2}}$ **h** $16^{-\frac{1}{4}}$ **i** $8^{\frac{2}{3}}$

   j $16^{\frac{3}{4}}$

4 **a** Write these numbers in standard form.

     i 500   **ii** 2140   **iii** 120 000   **iv** 895.3

   b Write these numbers as ordinary numbers.

     i 4.2×10^3 **ii** 5.02×10^5 **iii** 7×10^6 **iv** 5.125×10^2

5 Work these out without using a calculator, giving your answer in standard form.

   a $(3 \times 10^5) \times (2 \times 10^7)$ **b** $(8 \times 10^{10}) \div (4 \times 10^3)$

   c $(5 \times 10^6) \times (3 \times 10^4)$ **d** $(1.6 \times 10^5) \div (2 \times 10^2)$

1 The mass of the Earth is 6×10^{24} kg to 1 significant figure. The mass of the Sun is approximately 330 000 times the mass of the Earth.

  a Write 330 000 in standard form.

  b Without using a calculator, work out the mass of the Sun, giving your answer in standard form.

2 The population of Sweden is approximately 9×10^6 people.

  a Write this number given in standard form as an ordinary number.

Sweden has an area of approximately 450 000 km².

  b Write this number in standard form.

  c Work out the population density without using a calculator and giving your answer in standard form.

3 a Simplify

  i $a^2 \times a^5$ **ii** $\dfrac{x^3}{x^7}$

  iii $\dfrac{y^6}{y^3 \times y}$

  b Evaluate

  i $49^{\frac{1}{2}}$ **ii** $81^{-\frac{1}{4}}$

4 a Simplify $3a^2 b \times 4a^3 b^4$

  b Evaluate

  i $(3^2)^2$ **ii** $(\sqrt{7})^2$

  iii $\sqrt{(2^2 \times 3^2)}$

1 a Evaluate

    i $\frac{3}{4}$ of 72   **ii** 45% of 320   **iii** $\frac{4}{9}$ of 216   **iv** 24% of 825

  b Write these fractions as percentages.

    i $\frac{17}{25}$ **ii** $\frac{19}{20}$ **iii** $\frac{113}{200}$ **iv** $\frac{7}{15}$

2 a Find the nth term of these linear sequences.

    i 9, 11, 13, 15, 17, ...   **ii** 17, 14, 11, 8, 5, ...

    iii $3\frac{1}{4}$, $3\frac{1}{2}$, $3\frac{3}{4}$, 4, $4\frac{1}{4}$, ...

  b Find the nth term of this quadratic sequence.

    −2, −2, 0, 4, 10, 18, ...

3 a Rotate flag A 90° clockwise about the origin. Label the image B.

  b Reflect flag B in the line $x = 0$. Label the image C.

  c Describe fully the transformation that maps flag A onto flag C.

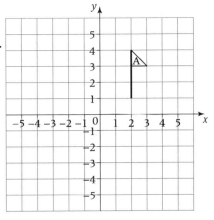

4 Adam chooses a starter and a main course from the menu at 'Chez Banks'. List all the possible ways that he could choose his meal. The list has been started for you.

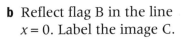

Starter	Main
Soup	Lasagne
Soup	Rack of lamb
Soup	Roast chicken
Soup	Beef en croute
Paté	

1 Copy and complete the table. Choose from the words *identity*, *equation* or *formula* for the right-hand column. The first one has been completed for you.

	Identity, equation or formula?
$a \times a \times a = a^3$	identity
$3x + 4 = 9x - 2$	
$4p^2(p - 1) = 4p^3 - 4p^2$	
$A = \pi r^2$	
$x^2 = 49$	
$V = lwh$	

2 Use the formula

$$\text{Area of a trapezium} = \tfrac{1}{2}(a + b)h$$

to find

a the area of a trapezium with parallel sides 5 cm and 11 cm that are 6 cm apart

b the height of a trapezium with parallel sides 4 cm and 10 cm and an area of 56 cm^2.

3 Use the formula

$$\text{Average speed} = \frac{\text{distance}}{\text{time}}$$

to find

a the speed of a car in km/h which travelled a distance of 40 km in 30 minutes.

b the distance covered by a car travelling at 60 km/h for 1 hour and 20 minutes.

4 Write a formula to represent these quantities.

a The amount, P, in pounds earned during one job by a plumber whose call out charge is £45 and whose hourly rate, h, is £30.

b The amount, C, that it costs in pounds to order a number of t-shirts, t, at £12 each from a mail order company whose postage and packing is £4, no matter how large or small the order.

1 Make x the subject of these formulae.

a $x - a = b$ b $px = q$ c $A = xy + r$

d $tx - v^2 = w^2$ e $P = x(q + r)$ f $ax^2 = b$

g $h = c(x - d)$ h $x^3 p = qr$ i $px - qx = n^2$

2 Lorne's father pays him 50p for every household chore, c, that he completes and £5 pocket money per week.

a Write a formula to show Lorne's total income, A, in pounds, for one week.

b Use your formula to work out how much money Lorne receives if he completes 7 chores in one week.

c Rearrange your formula to make c the subject.

d Use part **c** to work out how many chores Lorne must complete to earn £10 in one week.

3 Make y the subject of these formulae.

a $a - y = b$ b $p - qy = r$

c $v^2 - wy = x^2$ d $\dfrac{k}{y} = t$

e $h(b - y) = d$ f $\dfrac{1}{y} = \dfrac{p^2}{q^2}$

4 Kirsten is rearranging the formula

$$ax - b = p(x + q)$$

to make x the subject. The lines of her working have been muddled. Can you put them back in the correct order?

i $x(a - p) = pq + b$

ii $ax - px = pq + b$

iii $x = \dfrac{pq + b}{a - p}$

iv $ax - b = px + pq$

1 a Prove that the product of an odd number and an even number is always even.

 b Prove that squaring an even number will give a number that is divisible by 4.

Hint: You need to generalise to all possible examples to write a proof.

2 Lee uses this formula to work out values of a

$$A = 4.21b + 7.1$$

a Work out the value of A when $b = 2.56$.

Lee works out the value of A when $b = 3.15$. He gets 20.3615. Lee is correct but his friend, Oscar, argues that the answer is 43.1525.

b Can you explain where Oscar has made an error?

3 John said, 'When $x = 4$, then the value of $3x^2$ is 144'. Pat said, 'When $x = 4$, then the value of $3x^2$ is 48'.

a Who was right? Explain why.

b Work out the value of $3(x - 1)^2$ when $x = 4$.

4 Rearrange

$$p = q - \frac{k}{x}$$

to make x the subject.

5 Make y the subject of the formula

$$k(y - t) = xy - w$$

1 Write the answers to these calculations.

   a 0.3×0.4 **b** 0.06×0.5

   c 1.2×0.7 **d** $1.24 \div 0.4$

   e $3 \div 0.2$ **f** 1.05×0.5

   g $14.6 \div 0.4$ **h** $6.9 \div 0.03$

2 a Write the equation of the line that has a gradient of 3 and passes through $(0, 4)$.

   b Find the equation of the line that passes through $(1, 1)$ and $(3, -3)$.

3 Find the missing angles, giving reasons for your answers.

   a   **b**

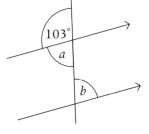
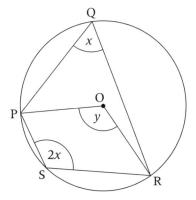

4 The table gives the results of 10 students in two papers of a Mathematics exam.

Paper 1	48	65	85	96	75	59	69	84	53	90
Paper 2	40	67	74	96	62	54	58	77	65	81

   a Represent these data on a scatter diagram.

   b Draw a line of best fit and use this to estimate a mark in paper 2 for a student who was absent on the day of the exam but who scored 73 on paper 1.

1 Which of these pairs of triangles are congruent? Give reasons for your answer.

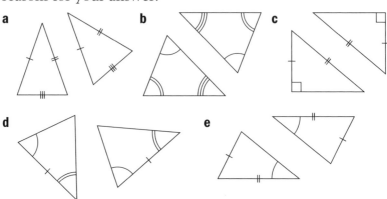

a **b** **c**

d **e**

2 In this diagram

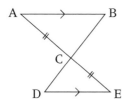

   a Show that angle DCE = angle ACB.

   b Show that angle BAC = angle DEC.

   c Prove that triangle ABC is congruent to triangle CDE.

3 PQRS is a kite.
   Angle PQR = 76°

   a Find angle PQT.

   b Find angle QPT.

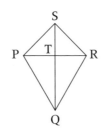

1 Pythagoras' theorem states that
$a^2 + b^2 = c^2$ for right-angled triangles.

A Pythagorean triple is a set of three
whole numbers that fit the rule
$a^2 + b^2 = c^2$. Two commonly used triples
are 3, 4, 5 and 5, 12, 13.

a Check that $a^2 + b^2 = c^2$ if $a = 3$, $b = 4$ and $c = 5$. Similarly
check when $a = 5$, $b = 12$ and $c = 13$.

b Copy and complete this table of Pythagorean triples.

a	b	c	Check
3	4	5	$9 + 16 = 25$
5	12	13	$25 + 144 = 169$
7	24		
	40	41	
11		61	

c Look at the sequence in column a. Look at the
relationship between b and c and compare this with a.
Predict the next two Pythagorean triples. Use $a^2 + b^2 = c^2$
to check.

Hint: $b + c = ?$

2 Work out the missing sides in these right-angled triangles.

a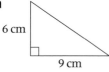
6 cm
9 cm

b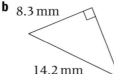
8.3 mm
14.2 mm

c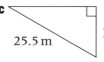
25.5 m
12.4 m

3 a Work out the length of the diagonal of a square of side 4 cm.

b Work out the length of the diagonal of a rectangle with
dimensions 7.5 cm and 3.5 cm.

1 Work out the area of this rhombus.

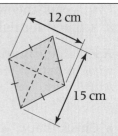

12 cm

15 cm

2 a Draw a set of axes from 0 to 8 in both the *x* and *y* directions. Plot the coordinates A = (1, 1), B = (4, 5) and C = (8, 3). Join these points to form triangle ABC.

b Use Pythagoras' theorem to work out the length of each side of this triangle.

3 Prove that this triangle is right-angled.

Hint: The converse of Pythagoras' theorem is true; that is, if the sides of the triangle are *a*, *b* and *c* such that $a^2 + b^2 = c^2$ then the triangle must be right-angled.

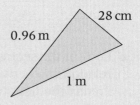

28 cm

0.96 m

1 m

4 The diagram shows a cylindrical pencil pot.
The diameter of the base is 10 cm and its height is 14 cm.
Emma has a pencil of length 18 cm. Can Emma place her pencil in the pot so that it does not show above the rim of the pot? Show all your working.

Hint: Emma cannot place the pencil in vertically as the height of the pot is only 14 cm. Use Pythagoras' theorem to work out the length of the diagonal in a triangle with base 10 cm and height 14 cm.

1 Evaluate each of these.

    a 100^0 **b** 500^1

    c $(2^2)^4$ **d** $(\sqrt{7})^2$

    e $36^{\frac{1}{2}}$ **f** 5^{-1}

    g $8^{-\frac{1}{3}}$ **h** $4^{\frac{3}{2}}$

2 Generate the first five terms of each of these sequences.

    a $T_n = 2n + 5$ **b** $T_n = 12 - n$

    c $T_n = n^2 + 2n + 3$ **d** $T_n = n^3 + 1$

    e $T_n = \dfrac{1}{n(n + 1)}$

3 a Work out the length of a diagonal of a rectangle with length 12.8 cm and width 9.2 cm. Give your answer to 3 significant figures.

    b Work out the width of a rectangle with length 24 mm and a diagonal length of 26 mm.

    Hint: Draw sketches to help you.

4 Claire recorded the length of time taken in seconds to complete a four-piece puzzle by each toddler in a group of 2-year-olds.

9	15	26	12	30	45	39	32	25
44	36	28	8	27	33	45	41	40
31	21	17						

    a Represent these data on a stem-and-leaf diagram.

    b Describe the trend in these data.

    c Calculate
       i the median time
      ii the interquartile range.

1 Use BIDMAS to work these out.

    a $8 \times 9 + 5$ **b** $12 \div 4 \times 3$

    c $12 \times (17 - 5)$ **d** $3 \times 5 + 6 \times 4$

    e $8 \times (9 - 3) \times 2$ **f** $6^2 + 15 \div 3$

    g $\dfrac{5 \times (4^2 - 2)}{7}$ **h** $\sqrt{120 - 2^3 \times 7}$

2 Copy these calculations, inserting brackets *if necessary*, to make the answers correct.

    a $12 \times 3 + 4 = 84$ **b** $24 \div 6 \times 2 = 8$

    c $5 \times 4 + 2 \times 7 = 210$ **d** $40 \div 4 + 2^2 = 5$

    e $5 \times 4^2 \div 8 = 10$ **f** $15^2 - 10 \times 5 = 175$

    g $\dfrac{8^2 \div 2^3 \times 4}{2} = 1$ **h** $\sqrt{150 - 7^2} - 4 \times 5 = 11$

3 Calculate these, leaving π in your answers.

    a the area of a circle of diameter 10 cm

    b the volume of a sphere of radius 6 m

Hint: Volume of sphere $= \frac{4}{3}\pi r^3$

    c the radius of a circle of circumference 12 mm

4 Evaluate these without using a calculator.

    a $\sqrt{3} \times \sqrt{3}$ **b** $4 + \sqrt{4} \times \sqrt{4}$ **c** $\sqrt{8} \times \sqrt{2}$ **d** $\sqrt{5} \times \sqrt{20}$

Hint: For parts **c** and **d** use the fact that $\sqrt{a} \times \sqrt{b} = \sqrt{a \times b}$.

5 Simplify these surds.

    a $\sqrt{12}$ **b** $\sqrt{18}$ **c** $\sqrt{45}$ **d** $\sqrt{63}$

Hint: Use the fact that $\sqrt{ab} = \sqrt{a \times b} = \sqrt{a} \times \sqrt{b}$.

1 By copying and completing this table, write whether performing the operation on a positive number will give a result that is *larger*, *smaller* or *the same size*.

	The result is larger, smaller or the same size?
× 1.4	larger
÷ 0.2	
× 0.8	
÷ 1.75	
× 1	

2 True or False?
   a Dividing by 0.5 is the same as multiplying by 2. Show your working.
   b Multiplying by 0.1 is the same as dividing by 10. Show your working.

   Hint: Use fractions to demonstrate.

3 Work these out without using a calculator.
   a 3×0.2 b 0.05×6 c $12 \div 0.3$
   d $21 \div 0.07$ e 0.4×0.15 f $4.5 \div 0.9$
   g 1.2×0.011 h $2.25 \div 0.015$

4 Use a written method to work out
   a $45.9 + 18.3$ b $32.56 - 18.37$ c $82.5 + 9.36$
   d $12.9 - 4.38$ e $2.08 + 0.589$ f $8.03 - 0.214$

5 Use a written method to work out
   a 41.2×5 b $22.75 \div 7$ c 15.6×0.4
   d $44.8 \div 0.8$ e 12.8×9.7 f $5.904 \div 0.18$

1 Given that $\sqrt{3} \approx 1.73$ and $\sqrt{5} \approx 2.24$, find approximate values for these surds, without using a calculator. Show full working.

   a $\sqrt{12}$ **b** $\sqrt{20}$ **c** $\sqrt{45}$ **d** $\sqrt{15}$

   Hint: Simplify each surd using the fact that $\sqrt{a \times b} = \sqrt{a} \times \sqrt{b}$

2 Given that $125 \times 46 = 5750$, evaluate

   a 1.25×4.6

   b 12.5×0.46

   c $57.5 \div 1.25$

   d $5.75 \div 0.46$

3 Work out the missing side of this right-angled triangle, leaving your answer in surd form.

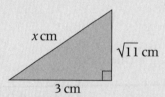

4 Calculate

$$\sqrt{0.5^2 + 1.4^2 - 2 \times 0.5 \times 1.4}$$

5 A metal box has a mass of 115 g measured to the nearest gram and a volume of 12.8 cm^3 measured to 1 decimal place. Work out the upper and lower bounds of the density of the metal used to make the box.

1 Evaluate these calculations.

    a $4 + 8 \times 3$ **b** $12 \div 2 + 4$ **c** $3 + 1 \times 8 + 4$

    d $5 \times (7 + 4)$ **e** 3×4^2 **f** $3^2 + 7 \times 4$

    g $(8 - 6)^3 \div 4$ **h** $\dfrac{3 \times (6^2 - 10)}{6}$

Hint: Use BIDMAS

2 a Without drawing the graphs, write where these pairs of lines intersect.

    i $x = 2$ and $y = 5$ **ii** $y = -4$ and $x = 3$ **iii** $x = \frac{2}{3}$ and $y = -1$

  b Copy and complete this table to draw the graph of $y = 3 - 2x$.

x	-2	0	2
y	7		

    Use axes labelled from -8 to $+8$ in both the x and y directions.

3 Use Pythagoras' theorem to work out the length of the line segment joining these points.

    a $(1, 4)$ and $(2, 8)$ **b** $(2, -1)$ and $(5, 5)$
    c $(-3, -2)$ and $(-1, 3)$

Hint: Sketch out each line.

4 A biased dice in the shape of a tetrahedron is thrown 100 times. The table shows the outcome of the experiment.

Score	1	2	3	4
Frequency	48	14	16	22

  a The dice is thrown again. Find the probability that it will land on a 2 or 3.

  b If the dice is thrown 500 times, how many times would you expect it to land on a 1?

1 Solve these double-sided equations.

  a $4(x+3) = 5(x+2)$ **b** $8(2y+1) = 3(7y-4)$

  c $5(1-p) = 3(2p+9)$ **d** $3(5-2q) = 7(1-2q)$

  e $5a + 2(7a-3) = 51$ **f** $4x - 3(10-2x) = 20$

  g $(x+2)(x+5) = (x+3)^2$ **h** $(y-3)^2 = (y+4)^2$

2 A square has side $2(x-1)$.
A rectangle has length $2x+1$ and width $x-1$.

 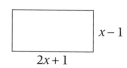

The perimeters of these shapes are equal.
What are their dimensions?

3 Solve these quadratic equations by factorising.

  a $x^2 + 6x + 8 = 0$ **b** $x^2 + 11x + 24 = 0$

  c $x^2 + 5x - 6 = 0$ **d** $x^2 + 6x - 16 = 0$

  e $x^2 = 9x - 20$ **f** $x^2 + 7 = 8x$

  g $x^2 - 4x = 0$ **h** $2x^2 + 12x = 0$

Hint: Parts **g** and **h** factorise into single brackets.

4 Two numbers have a difference of 5 and a product of 84.
Form a quadratic equation and solve to find the *two* sets of
solutions.

Hint: Let the numbers be x and $x+5$.

Trial and improvement and simultaneous equations

1 This length of this rectangle is 5 cm more than its width.
The area of this rectangle is 33.44 cm^2.
What are its length and width?

Write an equation to solve this problem.
Use trial and improvement to find the exact length and width of the rectangle.

2 Copy, extend and complete the table to solve

$$x^3 + x^2 = 180$$

to 2 decimal places.

x	$x^3 + x^2$	Comment
5.3	176.967	Too small
5.4	186.624	Too big

Hint: Look at 3 decimal places in order to round to 2 decimal places.

3 Solve these simultaneous equations by either adding or subtracting to eliminate one variable.

a $x + 2y = 7$
$x + y = 4$

b $5x + 2y = 8$
$2x + 2y = 2$

c $p - 3q = 10$
$4p + 3q = 10$

d $a = 5b + 8$
$3b = 8 - a$

4 Here are my receipts from two visits to 'Café Connection'.
By setting up a pair of simultaneous equations work out the cost of a latte.

2 Lattes
3 Carrot Cakes
£5.85

1 Carrot Cake
2 Lattes
£3.95

1 Solve these simultaneous equations by the elimination method.

a $3x + y = 11$
$2x + 2y = 10$

b $4x - y = 9$
$3x + 2y = 4$

c $5a + 3b = 19$
$3a - 2b = 19$

2 Solve these equations.

a $7a - 3 = 5a + 7$

b $6(b + 2) = 54$

c $5c + 4 = 2(c - 3)$

3 a Expand these

  i $(5x - 2)(x + 2)$

  ii $(2x - 5)^2$

b Solve the equation
$x^2 + 5x - 24 = 0$

4 A cuboid has dimensions as shown in the diagram.
The volume of the cuboid is 70 cm^2.

a Show that $x^3 + x^2 = 70$

b The equation $x^3 + x^2 = 70$ has a solution between $x = 3$ and $x = 4$.
Use trial and improvement to find this solution correct to 1 decimal place.

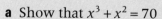

$x + 1$

x

x

Hint: Use a table to show all your working.

1 Copy these calculations, inserting brackets *if necessary*, to make the answers correct.

a $8 + 7 \times 3 = 29$

b $12 \div 5 - 1 = 3$

c $3 \times 9 - 2 \times 4 = 84$

d $12^2 \times 5 - 3 = 288$

e $\sqrt{24 \div 4 - 2} = 2$

2 Solve these quadratic equations by factorising.

a $x^2 + 10x + 21 = 0$

b $a^2 + 4a - 45 = 0$

c $b^2 - 5b - 36 = 0$

d $y^2 - 5y = 0$

e $3t^2 + 12t = 0$

3 Which of these pairs of triangles are congruent?

a 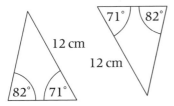 **b**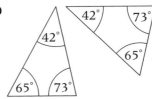

Explain your answer.

4 A group of children are asked to choose their favourite primary colour. The table gives this information.

	Red	Blue	Yellow	Total
Girls	5		5	
Boys		8		16
Total	11	12		

a Copy and complete the table.

b A child is chosen at random. Find the probability that the child

i prefers blue **ii** is a boy who prefers red.

c If these children are a sample from a school of 210 children, how many of the larger group would you expect to prefer yellow?

1 Two dice are thrown and their scores added together. The results of 100 throws are shown in the table.

Score	2	3	4	5	6	7	8	9	10	11	12
Frequency	4	4	7	13	10	20	16	12	5	7	2

a Work out the mean score per throw.

b Work out the

   i median   **ii** interquartile range.

2 Five coins are tossed and the number of heads is recorded each time. The results of 200 tosses are shown in the table.

No. of heads	0	1	2	3	4	5
Frequency	10	34	56	65	30	5

a Work out the

   i mean

   ii median

   iii modal number of heads per toss.

b Work out the range.

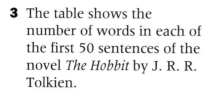

3 The table shows the number of words in each of the first 50 sentences of the novel *The Hobbit* by J. R. R. Tolkien.

a Calculate an estimate for the mean number of words per sentence.

b Write the modal class.

c Write the class interval that contains the median.

No. of words in sentence	Frequency
1–10	13
11–20	10
21–30	9
31–40	7
41–50	8
51–60	2
61–70	1

1 A sample of 30 seedlings is taken from a pot of cress and their heights measured. The results are given in centimetres, taken to the nearest millimetre.

7.4	6.2	5.3	3.4	2.9	3.6	6.4	5.9	5.5	2.1
4.5	5.7	4.3	3.8	5.4	6.2	4.6	4.8	5.3	4.2
5.1	6.0	4.1	2.3	1.5	3.4	5.2	7.0	4.2	1.9

  a Using class intervals of $0 < h \leqslant 2$, $2 < h \leqslant 4$, $4 < h \leqslant 6$, $6 < h \leqslant 8$ where h is the height of the seedling in centimetres, construct a grouped frequency table for these data.

  b Draw a frequency polygon for these data.

2 A small mail order company kept a record of the length of phone calls it received during December and January.

December

Length of call m minutes	Frequency
$0 < m \leqslant 2$	0
$2 < m \leqslant 4$	32
$4 < m \leqslant 6$	54
$6 < m \leqslant 8$	131
$8 < m \leqslant 10$	83

January

Length of call m minutes	Frequency
$0 < m \leqslant 2$	44
$2 < m \leqslant 4$	55
$4 < m \leqslant 6$	27
$6 < m \leqslant 8$	14
$8 < m \leqslant 10$	10

  a Draw frequency polygons for these data.

  b For each set of data work out

    i the modal class   **ii** the range.

  c Use your answers to make comparisons, with reasons, between the length of calls received in December and January.

3 The owner of a bed-and-breakfast business kept a record of the takings in pounds per month over a period of 12 months.

Jan	Feb	Mar	Apr	May	Jun	Jul	Aug	Sep	Oct	Nov	Dec
630	630	665	770	945	1225	1400	1505	1120	875	700	840

Draw a time series graph to represent these data.

1 The table shows the number of letters in the first 100 words of the novel *Pride and Prejudice* by Jane Austen.

Word length	1	2	3	4	5	6	7	8	9	10	11	12	13
Frequency	6	28	19	17	10	4	3	5	1	2	3	1	1

Find the mean length of a word.

2 The table shows the number of hardback books sold in a supermarket each month from July to December.

Jul	Aug	Sep	Oct	Nov	Dec
185	250	162	155	301	396

a Draw a time series graph to represent these data.
b Work out the three-month moving averages for these data.

3 A furniture company keeps a record of its sales in one week. The table gives information about those sales which are £500 or less.

Cost, £c	Frequency
$0 < c \leqslant 100$	3
$100 < c \leqslant 200$	10
$200 < c \leqslant 300$	18
$300 < c \leqslant 400$	12
$400 < c \leqslant 500$	7

a Calculate an estimate for the mean from this table.
b Find the class interval in which the median lies.
c There was only one other sale this week, not included in the table. This sale cost £1500.
The manager says, 'The class interval in which the median lies will change.' Is the manager correct? Explain your answer.

1 Simplify these quantities.

a $\sqrt{3} + \sqrt{3}$ **b** $\sqrt{5} \times \sqrt{5}$

c $4\sqrt{2} + \sqrt{2}$ **d** $\sqrt{20}$

e $(4 + \sqrt{7})(4 - \sqrt{7})$ **f** $(5 + \sqrt{2})(5 + \sqrt{2})$

2 Rearrange these formulae to make x the subject.

a $a + x = b$ **b** $xy = z$

c $px + q = r^2$ **d** $x^2(y - p) = q$

e $t - x = w$ **f** $\dfrac{x}{ab} = c$

g $\dfrac{a}{x} = p - q$ **h** $r(h - x) = m$

3 A square has diagonals 20 cm long.
Use a sketch diagram to help you

a work out the area of the square

b work out the length of a side of the square.

Hint: Look at the square as two triangles.

Hint: Use square roots.

4 Doris did a survey of the number of occupants of all the cars passing her house during one hour. Her results are shown in the table.

No. of occupants	1	2	3	4
Frequency	17	13	9	11

a Work out the mean number of occupants per car.

b What is the modal number of occupants?

1 a Draw a diagram to show the position of the points A and B where the bearing of B from A is 112°.

b What is the bearing of A from B?

2 Mandy's house is 8 km due east of Dave's house. Mandy walks 4 km from her house on a bearing of 300° to school. After school she walks directly to Dave's house.

a Draw a diagram using a scale of 1 cm to 1 km to represent this information.

b By measuring, work out the distance from the school to Dave's house.

3 Y is due north of X. The bearing of Z from Y is 124°. The bearing of X from Z is 233°.

a Draw a sketch diagram to represent this information.

b Work out the three angles of the triangle XYZ.

c What is the bearing of Z from X?

4 a Construct a triangle with sides of length 6 cm, 8 cm and 10 cm.

b What type of triangle have you drawn?

c Check your assumption in part **b** using an algebraic method.

5 True or False? It is possible to construct a triangle with sides of length 5 cm, 6 cm and 11 cm.
Explain your answer *without* drawing the triangle.

1 **a** Draw a line 6 cm long and label it AB. Construct the perpendicular bisector of AB and label it CD. Label the point where these two lines meet as X.

   b Bisect the angle CXB. Label this new line XY.

   c What is the size of angle CXY?

2 **a** Construct an angle of 60°.

   b Bisect this angle to give an angle of 30°.

   Hint: This is an angle of an equilateral triangle.

3 **a** Draw a circle of radius 4 cm.

   b Without altering the compasses, place the point of the compasses on the circumference of the circle and draw an arc somewhere else on the circumference.

   c Again, without altering your compasses, place the point of the compasses where the arc cuts the circumference and draw a second arc. Repeat this process until you arrive back at the first arc.

   d Join up each point on the circumference that is cut by an arc to an adjacent point on the circumference that is cut by an arc.

   e Name the shape that you have constructed.

4 **a** Trace this line and points marked *x* and *y*.

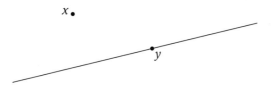

   b Use a ruler and compasses to construct a perpendicular from the point *x to* the line.

   c Use a ruler and compasses to construct a perpendicular from the point *y on* the line.

1 a A and B are two points 8 cm apart. Draw the locus of points that are equidistant from A and B.

b On the same diagram, draw the locus of points that are 3 cm from A.

2 a Copy the diagram of a garden ABCD using a scale of 1 cm to 1 m.

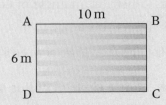

b A sprinkler is placed at point A. The sprinkler waters the garden up to 5 m away as it rotates.
A second sprinkler is placed at point C. It has a maximum reach of 4 m. Shade the area of the garden that remains unwatered.

3 a Construct an isosceles triangle with sides 6 cm, 6 cm and 4 cm.

b A garden is in the shape of an isosceles triangle as shown.

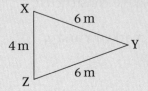

A water feature is to be placed in the garden so that it is

• nearer to YZ than XY
• closer to point Y than to point Z.

Use your diagram from part **a** to represent the area where the water feature may be placed. Shade this area.

1 Work out these, giving your answer in index form.

a $5^2 \times 5^4$ **b** 3×3^2 **c** $a^5 \times a^6$

d $8^3 \div 8^3$ **e** $b^3 \div b^8$ **f** $4^3 \div 4^2$

g $7^2 \times 7^4 \times 7^6$ **h** $2^7 \times 2^9 \div 2^5$

2 This sequence does not start with term 1. Find the nth term of the sequence.

    12th term = 67,   13th term = 72,   14th term = 77

3 **a** Draw a sketch diagram to show the position of the points A and B where the bearing of B from A is 110°.

b What is the bearing of A from B?

c Point C is due south of B and the bearing of A from C is 330°. Add point C to your diagram and work out the angles of the triangle ABC.

4 The tables show the amount spent, in pounds, of the first 100 people to visit outdoor equipment stores on one day in Fort William, Scotland and Scunthorpe, Lincolnshire.

Fort William, Scotland

Amount (a)	Frequency
$0 \leqslant a < 20$	12
$20 \leqslant a < 40$	15
$40 \leqslant a < 60$	22
$60 \leqslant a < 80$	33
$80 \leqslant a < 100$	18

Scunthorpe, Lincolnshire

Amount (a)	Frequency
$0 \leqslant a < 20$	17
$20 \leqslant a < 40$	42
$40 \leqslant a < 60$	28
$60 \leqslant a < 80$	10
$80 \leqslant a < 100$	3

a Draw frequency polygons for these data.

b Use your frequency polygons to make comparisons, with reasons, between the amounts spent in the two outdoor equipment stores. Try to include the modal class and the range in your descriptions.

Do not use a calculator for this exercise.

1 Work out these quantities, giving your answers as fractions or mixed numbers where appropriate and showing any cancelling.

a $\frac{1}{3}$ of 105 m

b $\frac{1}{7}$ of 252 g

c $\frac{3}{4}$ of 52 cm

d $\frac{2}{5}$ of 55 litres

e $\frac{5}{8}$ of 96 kg

f $\frac{7}{12}$ of 126 mm

g $\frac{4}{9}$ of 21 km

h $\frac{4}{15}$ of 63 cl

2 A charity sells plastic bracelets advertising their cause for £2. Of each bracelet sold, $\frac{3}{10}$ of the money is used to pay overheads and $\frac{1}{8}$ is used for marketing. The rest of the money is used directly for charity work.

a How much money from each bracelet is used to pay overheads?

b How much money is used directly for charity work?

3 Work out these quantities using an appropriate written method.

a 5% of £540

b 15% of $620

c 12% of €625

d 24% of £325

e 52% of £275

f 73% of $1400

g 64% of €350

h 87% of £2600

Hint: Find 10% by dividing the number by 10 and/or 1% by dividing the number by 100.

4 True or False?
25% of $\frac{4}{5}$ of 960 is greater than 45% of $\frac{2}{5}$ of 960.

1 If a quantity is to be *in*creased by 20% we multiply by 1.2. Similarly, if a quantity is to be *de*creased by 20% we multiply by 0.8. Copy and complete the table.

Percentage change	Multiply by ...
Increase of 5%	
Decrease of 12%	
	1.75
	0.45
Increase of 150%	

2 Use a calculator to find each of these percentage changes.

a Increase £36 by 22%   **b** Increase 15 m by 24%

c Decrease $42 by 54%   **d** Increase 65 kg by 12%

e Decrease 256 litres by 8%   **f** Decrease 952 km by 35%

3 a A winter coat costs £200. In the January sale, the coat is reduced by 15%. Work out the cost of the coat in the sale.

b In the last week of the sale, the coat is reduced by a further 20%. What is the cost of the coat now?

c By what single decimal number can £200 be multiplied in order to directly find the cost of the coat in the last week of the sale?

Hint: £200 × x is the same as £200 × 0.85 × 0.8. What is x?

d Hence work out the overall percentage reduction applied to the cost of the coat.

Hint: It is *not* 35%.

4 Mac wants to invest £500 for 5 years. His bank offers him two options.

Option 1 is simple interest of 4.25% per annum

Option 2 is compound interest of 4% per annum

Which option should Mac choose in order to achieve the most interest on his investment? Show your working.

1 There are 1200 students at Cornerhouse School.
55% of these students are girls.

   a Work out the number of girls at Cornerhouse School.

   There are 168 students in Year 11.

   b Work out the percentage of students at Cornerhouse School that are in Year 11.

2 Work out the cost of this television before the sale.

SALE! 12% off
23 inch flat screen
LCD television
Now ONLY £699.60

3 A Geography test has two parts, part 1 and part 2.
Part 1 is out of 25 marks.
Part 2 is out of 40 marks.
Jimmy scored a total of 52 marks for the test.
Lynda scored 92% in Part 1 of the test and 80% in Part 2.
Who scored the highest mark on the test?
Show all your working.

4 Niamh spent $\frac{1}{8}$ of her birthday money on a CD and $\frac{1}{3}$ of her birthday money on clothes. She saved the rest.
Work out the fraction of money that Niamh saved.

1 Calculate the missing side x, leaving surds in your answer.

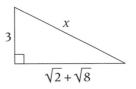

Hint: Recall that $(\sqrt{2} + \sqrt{8})^2 = (\sqrt{2} + \sqrt{8})(\sqrt{2} + \sqrt{8})$. Multiply out this quadratic remembering to do 'inners', 'outers', 'firsts' and 'lasts'.

2 Solve these simultaneous equations algebraically.

a $2p + 3q = 12$
$p - q = 1$

b $5a + 3b = 7$
$2a - 2b = 6$

c $4x - y = 19$
$2x - 3y = 7$

3 a Construct a triangle with a base of length 4 cm and sides of length 5 cm and 5 cm.

b What is the name of this type of triangle?

c Construct a perpendicular from a vertex to the base of 4 cm. Hence find the perpendicular height of the triangle using Pythagoras' theorem. Measure this distance on your diagram and check the accuracy of your construction.

4 A father-to-be runs a competition for his friends where they must correctly guess the mass of his baby at birth in order to win a prize. The results of the guesses of his 25 friends are given in kg.

2.5	3.45	3.25	4	3.75	2.8	4.15	3.9	3.95
2.95	2.85	3.1	3.6	4.1	3.5	3.55	3.7	2.9
3.25	3.2	3.9	4	4.05	3.75	3.5		

a Using class intervals of $2 < m \leqslant 2.5$, $2.5 < m \leqslant 3$, etc. where m is the mass of the baby in kg, construct a grouped frequency table for these data.

b Draw a frequency polygon for these data.

1 The monthly salary of a sample of 100 office workers is recorded in the table.

Salary, £S	Frequency, f	S	Cumulative frequency
$500 < S \leqslant 750$	12	$\leqslant 750$	12
$750 < S \leqslant 1000$	22	$\leqslant 1000$	34
$1000 < S \leqslant 1250$	36		
$1250 < S \leqslant 1500$	18		
$1500 < S \leqslant 1750$	8		
$1750 < S \leqslant 2000$	4		

a Copy and complete the cumulative frequency table.

b Draw a cumulative frequency diagram for these data.

c Write the modal class interval.

2 A Sunday league football team records the number of minutes that pass before the first goal is scored in each match that they play.

Number of minutes, m	Frequency
$0 < m \leqslant 15$	5
$15 < m \leqslant 30$	9
$30 < m \leqslant 45$	12
$45 < m \leqslant 60$	8
$60 < m \leqslant 75$	3
$75 < m \leqslant 90$	5

a Draw a cumulative frequency table for these data.

b Draw a cumulative frequency diagram for these data.

c Estimate (from the diagram)

   i the median   **ii** the interquartile range .

3 The table gives the exam results of a group of 120 students.

Test result, $t\%$	Frequency
$40 < t \leqslant 50$	9
$50 < t \leqslant 60$	18
$60 < t \leqslant 70$	28
$70 < t \leqslant 80$	44
$80 < t \leqslant 90$	15
$90 < t \leqslant 100$	6

a Draw a cumulative frequency table and diagram for these data.

b Estimate (from the diagram)

   i the median

   ii the interquartile range.

c Estimate the number of students that passed the exam if the pass mark was 55%.

1 The diagram shows the cumulative frequency of the yield from 50 of each of two types of tomato plant.
Write three comparisons concerning the yield (number of tomatoes).

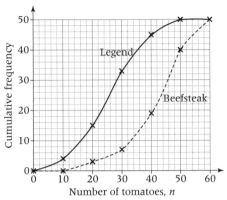

2 Use the graph above to find

   a the median for each variety

   b the lower and upper quartiles for each variety.

   c Use your results to draw two box plots.

   Hint: Use lower bound of first class and upper bound of last class to estimate the minimum and maximum, as they are not given.

3 The tables record the mass, in grams, of 100 tomatoes from two different varieties, Legend and Beefsteak.

Legend

Mass, m (g)	Frequency
$200 < m \leqslant 210$	15
$210 < m \leqslant 220$	22
$220 < m \leqslant 230$	40
$230 < m \leqslant 240$	14
$240 < m \leqslant 250$	9

Beefsteak

Mass, m (g)	Frequency
$200 < m \leqslant 210$	8
$210 < m \leqslant 220$	12
$220 < m \leqslant 230$	19
$230 < m \leqslant 240$	36
$240 < m \leqslant 250$	25

   a Copy the tables and add an extra column entitled 'Cumulative frequency'. Calculate the cumulative frequencies.

   b Draw a cumulative frequency diagram for each variety.

   c Find (for each variety of tomato from your diagrams)

     i the median

     ii the lower and upper quartiles.

   d Use your results to draw two box plots.

1 The cumulative frequency diagram gives information about the test results of some students.

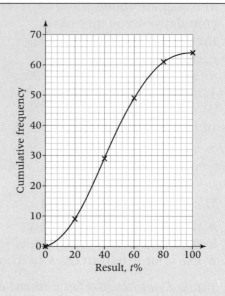

  a How many students were in the group?

  b Use the diagram to estimate the median test result.

  c If the pass mark was 55%, use the diagram to estimate the number of students who passed the test.

2 Use the information from the diagram in question 1 to draw a box plot for these data. The minimum mark attained in the test was 12% and the maximum mark was 88%.

3 The cumulative frequency diagram shows the time taken for 50 girls to complete a puzzle.

The maximum time was 52 seconds and the minimum time was 9 seconds.

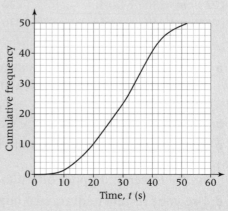

  a Use the diagram to estimate the median time taken.

  b Draw a box plot showing information about the girls' times.

1 a Julia bought her house for £120 000 in 1999. Two years later, an estate agent valued the house and deduced that the cost had risen by 34%. Work out the value of Julia's house in 2001.

  b In 2004, Julia decided to sell her house and placed it on the market at £276 000. What was the percentage increase of the house price in the five years that Julia owned it?

2 Solve these quadratic equations.

  a $x^2 + 7x + 12 = 0$ **b** $x^2 + 9x + 14 = 0$

  c $x^2 + 4x - 5 = 0$ **d** $x^2 - 14x + 40 = 0$

  e $x^2 = 5x - 24$ **f** $x^2 - 9x = 0$

3 A point, P, moves so that it is always 2 cm away from the line AB. Copy the line AB and draw the locus of P.

A  B

4 Ryan constructed a table detailing the amount he spent per quarter on gas.

	Jan–Mar	Apr–Jun	Jul–Sep	Oct–Dec
2004	£80	£35	£28	£75
2005	£84	£38	£34	£76

  a Draw a time series graph to represent these data.

  b Calculate four-point moving averages for these data and plot these on the same graph.

  c Describe the trend shown by the moving averages.

1 For this solid, draw

  a the plan

  b the front elevation

  c the side elevation.

2 These diagrams show the plan and front elevation of a solid.

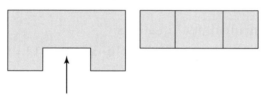

The arrow shows the direction from which the front elevation was drawn. Sketch the solid.

3 Work out the volume of this prism.

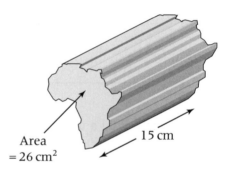

Area = 26 cm²

15 cm

4 Work out the volume of these prisms.

  a

  b

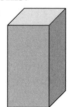

Radius of base = 6 cm
Height = 10 cm

Length = 4.2 m,
width = 3.7 m and
height = 8 m

1 a Work out the volume of a cube with an edge of length 7 cm.

   b Work out the volume of a cube whose surface area is 96 cm².

   Hint: Find the area of one face. Then find the length of one edge.

2 Work out the surface area of this triangular prism which has a cross-section in the shape of an equilateral triangle of side 6 cm and a length of 12 cm.

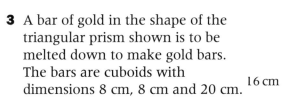

3 A bar of gold in the shape of the triangular prism shown is to be melted down to make gold bars. The bars are cuboids with dimensions 8 cm, 8 cm and 20 cm.

   How many gold bars can be made?

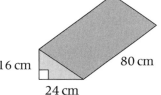

16 cm   80 cm
24 cm

4 Change these measurements to the units given.

   a 3500 cm to m   **b** 184 m to cm

   c 0.01 m to mm   **d** 0.05 km to cm

   e 15 m² to cm²   **f** 8400 mm² to cm²

   g 2.5 cm³ to mm³   **h** 800 000 cm³ to m³

5 If a, b, c, r and h represent lengths, use dimension theory to deduce whether each of these expressions is a length, area, volume or none of these.

   a abc **b** $\frac{2}{3}bc$ **c** $3bh + \pi r$ **d** $5h^2 a$

   e $a + \pi b$ **f** $ab^2 + r^2$ **g** $a(b + 2r)$ **h** $\frac{1}{2}\pi r^2 h$

1 Billy wrote down the volume of a cylinder to be

$V = 2\pi rh$

Shirley was not sure of the correct answer but told Billy that he was wrong. How did she explain her answer?

2 A cube has a surface area of 150 cm^2.
Work out the volume of the cube.

3 Change 9.5 m^2 to cm^2.

4 A car park token is in the shape of a cylinder with diameter 2 cm and a thickness of 0.4 cm.
The token is made out of brass which has a density of 8.6 g/cm^3.
Work out the mass of the token.

5 Caroline sets off from home for work at 7.45 a.m. and arrives at 8.10 a.m. She drives at an average speed of 56 km/h.
How far does Caroline travel to work?

1 This is a list of ingredients for making fish-cakes for 4 people.

Work out the amount of ingredients needed to make fish-cakes for 10 people.

> 150 g cod
> 100 g wild Alaskan salmon
> 250 g mashed potatoes
> 1 egg
> 80 g breadcrumbs

2 Solve these simultaneous equations algebraically.

   a $x + y = 6$ **b** $2x + y = 3$ **c** $2x - 3y = 2$

       $2x - y = 9$ $3x + 3y = 3$ $5x + 2y = 24$

3 **a** Draw a line 6 cm long. Bisect this line using a pair of compasses.

   b Using a pair of compasses, accurately draw a triangle with sides 4 cm, 6 cm and 7 cm.

4 Lois compiled a table of information about the length of time it took each of her classmates to solve a crossword puzzle.

Time, t (minutes)	Frequency	Midpoint	Midpoint × Frequency
$5 \leqslant t < 10$	3	7.5	$7.5 \times 3 = 22.5$
$10 \leqslant t < 15$	8		
$15 \leqslant t < 20$	12		
$20 \leqslant t < 25$	5		
$25 \leqslant t < 30$	2		

   a Copy and complete the table to find an estimate for the mean time taken in minutes and seconds.

   b Find

      i the modal class

      ii the class containing the median.

1 a Copy and complete this table to generate coordinates for the graph of $y = x^2 + 2x$.

x	-3	-2	-1	0	1	2	3
x^2		4					
$2x$		-4					
y		0					

b By drawing appropriate axes, plot the coordinates that you have found in part **a** and join them to form a smooth parabola.

c Write the coordinates of the minimum point of this parabola.

2 The point (2, 9) lies on which of these quadratic graphs?

$y = x^2 - 2x + 1$ $y = x^2 - 3x - 10$ $y = 2x^2 + 1$

$y = 3x^2 - 2x - 1$ $y = x^2 + 4x - 3$

3 a Copy and complete this table to generate coordinates for the graph of $y = x^3 + 2x^2$.

x	-3	-2	-1	0	1	2
x^3	-27					
$2x^2$	18					
y	-9					

b Draw and label an x-axis from -3 to 2 and a y-axis from -10 to 20. Plot the coordinates from the table in part **a**.

c Write the coordinates of the turning points of this cubic graph.

Hint: The turning points are literally the points where the graph turns a corner and changes direction. The gradient will change from positive to negative or negative to positive.

1 a Copy and complete the table to generate three coordinates for the line graph $y = 3x - 1$.

x	−1	0	1
y		−1	

b Draw an x-axis from −2 to 2 and a y-axis from −5 to 5. Plot the graph of $y = 3x - 1$ on your axes.

c Plot the graph of $y = 3 - x$ on the same set of axes.

d Use your graphs to solve the simultaneous equations

$$y = 3x - 1$$
$$y = 3 - x$$

2 Solve this pair of simultaneous equations graphically.

$$x - y = 2$$
$$y = 3x - 4$$

Hint: Rearrange $x - y = 2$ to make y the subject. Use question 1.

3 Explain why the simultaneous equations $y = 4x - 1$ and $y - 4x = 3$ have no solutions.

Hint: Use graphs to help you *or* look at the gradient of each line.

4 a Copy and complete this table to generate coordinates for the graph of $y = x^2 + 3x - 4$.

x	−5	−4	−3	−2	−1	0	1	2
x^2			9					
$3x$			−9					
−4			−4					
y			−4					

b By adding lines of your choice on the same set of axes solve graphically

i $x^2 + 3x - 4 = 0$ **ii** $x^2 + 3x - 4 = -6$

iii $x^2 + 3x - 4 = 3$ **iv** $x^2 + 3x - 4 = 1 - x$

v $x^2 + 3x - 4 = x - 5$

1 Plot the curve of $y = x^2 - x - 2$ for $-2 \leqslant x \leqslant 3$ and find the coordinates of its minimum point.

Hint: Read off the x-coordinate for the minimum point. Substitute this value into $y = x^2 - x - y^2$ to check the value of the y-coordinate.

2 This is a sketch of the graph $y = x^3 - x^2 - 6x$.

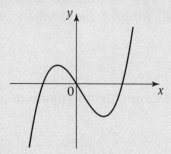

Find the coordinates of the points where the graph cuts the x-axis.

Hint: Fully factorise $x^3 - x^2 - 6x$. Take out a factor of x to begin. Factorise the quadratic that remains.

3 Given that the graph $y = x^2 - x - 6$ is already drawn, which *one* line would you need to add to the same axes to solve

a $x^2 - x - 6 = 5$ **b** $x^2 - x - 6 = 0$ **c** $x^2 - x - 6 = 3 - x$

d $x^2 - x = 4$ **e** $x^2 - 3x - 7 = 10$ **f** $x^2 + x = 10$?

4 Use a graphical method to solve this problem.

The sum of two numbers is 5. If I subtract one of these numbers from twice the other number, I get an answer of 7. What are my numbers?

Hint: Form a pair of simultaneous equations and plot their graphs.

1 A rectangle has dimensions of 5 cm and 3 cm, to the nearest cm. Find the upper and lower bound for the area of the rectangle.

2 a Copy and complete this table to generate coordinates for the graph of $y = x^2 - 2x - 3$.

x	-2	-1	0	1	2	3	4
x^2		1					
$-2x$		2					
-3		-3					
y		0					

b By adding lines of your choice on to the same set of axes solve graphically

  i $x^2 - 2x - 3 = 0$ **ii** $x^2 - 2x - 3 = -3$

  iii $x^2 - 2x - 3 = 1$ **iv** $x^2 - 2x - 3 = x + 1$

3 These diagrams show the plan and front elevation of a solid.

The arrow shows the direction from which the front elevation was drawn. Sketch the solid.

4 Harry recorded the amount of time, in hours, that he slept for each night in September. His results are shown in the table.

Time, t (hours)	Frequency
$0 < t \leqslant 3$	1
$3 < t \leqslant 6$	2
$6 < t \leqslant 9$	18
$9 < t \leqslant 12$	9

a Write the class interval that contains the median.

b Calculate an estimate for the mean amount of time Harry spent asleep.

1 Simplify these ratios.

  a £8 : £4

  b 50 kg : 5 kg

  c 2m : 20 000 m

  d 54p : 36p

  e 21 cm : 35 cm : 14 cm

  f £3 : 60p

  g 1.25 kg : 750 g

  h 625 m : 1.4 km

  i 25 min : 1.25 hours

  j 1.2 m : 45 cm : 2.25 m

2 Simplify these ratios.

  a $\frac{1}{2} : 2$ **b** $3 : \frac{1}{4}$ **c** $\frac{1}{2} : \frac{1}{5}$ **d** $\frac{2}{3} : \frac{4}{7}$

3 Write these ratios in the form $1 : n$.

  a 3 : 5

  b 15 : 4

  c 2 cm : 10 km

  d 5 cm : 2 km

4 Divide £480 in these ratios.

  a 2 : 1   **b** 1 : 5   **c** 5 : 3   **d** 7 : 5

5 Kevin and Kathleen decide to play the National Lottery. Kevin contributes £10.50 and Kathleen contributes £7.00 to buy tickets.

  a Write the ratio of Kevin's contribution to Kathleen's contribution in its simplest form.

Kevin and Kathleen win £1500.

  b Divide their prize money between them in the ratio of their original contributions.

1 A ratio can be written as a fraction. For example, the ratio $3 : 4$ can be written as the fraction $\frac{3}{4}$. Write these ratios as fractions, giving your answers in their simplest form where necessary.

  a $12 : 15$ **b** $24 : 64$ **c** $x : 3$ **d** $7 : y$

You can solve problems involving equivalent ratios using the methods for equivalent fractions.

Example

If $x : 4 = 2 : 5$ then	$\dfrac{x}{4} = \dfrac{2}{5}$
Multiplying through by 4	$x = \dfrac{8}{5} = 1\dfrac{3}{5}$

2 Find the missing values in these equivalent ratios.

  a $x : 5 = 9 : 15$ **b** $y : 3 = 1 : 2$ **c** $4 : a = 5 : 9$

3 A map has a scale of 1 : 50 000. Two villages are 6 cm apart on the map. How far apart are the actual villages? Give your answer in kilometres.

4 Archie, aged 9, Susie, aged 8, and Josie, aged 5, share a gift of £550 from a generous aunt in the ratio of their ages.

  a How much do they each receive?

Exactly a year later, they receive another gift of £550 from the same aunt. They decide to share the gift in the ratio of their ages again.

  b How much do they now each receive?

  c If this arrangement continues at the same time of year for every year of their lives, what happens to the amount each person receives as they grow older?

1 In a class, the ratio of girls to boys is 4 : 3. There are 16 girls in this class. Work out the number of boys.

2 a Lois, Sam and Alice have savings in the ratio 4 : 2 : 1. If Sam has £250, calculate the amount of money that Lois and Alice have in savings.

 b Lois, Sam and Alice are given £560 as a gift and decide to divide this between themselves in the ratio 4 : 2 : 1. Work out how much each person is able to add to their savings.

3 A bank pays compound interest of 4%. Find the amount of money in an account after 3 years if the original investment is £5000.

4 Gwen buys a new pair of shoes from her favourite shop. Gwen has a 20% discount card cut out from a magazine and she hands this over with her purchase. She pays £44 for the shoes.

What was the price of the shoes *before* the discount was applied?

5 In 2001, the United Kingdom had a population of 58.8 million. If the population of the United Kingdom is increasing at, on average, an annual rate of 0.4%, calculate an estimate for the population of the United Kingdom in 2005. Give your answer to 3 significant figures.

1 Write the upper and lower bounds of each of these measurements given that they have been measured to the degree of accuracy in brackets.

  a 12 m (to the nearest m)
  b 14.8 s (to 1 decimal place)
  c 1200 g (to 2 significant figures)
  d 24.5 kg (to 3 significant figures)
  e 3.05 litres (to 2 decimal places)
  f 4.39 m (to the nearest cm)

2 a Copy and complete this table to generate coordinates for the graph of $y = x^2 - x - 4$.

x	-2	-1	0	1	2	3	4
x^2	4						
$-x$	2						
-4	-4						
y	2						

  b By drawing appropriate axes, plot the coordinates that you have found in part **a** and join them to form a smooth parabola.
  c Write the coordinates of the minimum point of this parabola.

3 The cross-section of this prism is an isosceles triangle with a base of 10 cm and a *slant* height of 13 cm. The prism has a length of 30 cm. Work out the volume of the prism.

Hint: Use Pythagoras' theorem to find the perpendicular height of the triangular cross-section.

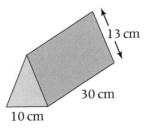

13 cm

30 cm

10 cm

4 The probability of rain in Derby on Tuesday is 0.65. What is the probability that Tuesday in Derby will be dry?

1 Copy this diagram.

  a Enlarge rectangle A by scale factor 3, centre (0, 0). Label the image B.

  b Write the perimeter of

    i rectangle A

    ii rectangle B.

  c How many times larger is the perimeter of rectangle B compared to the perimeter of rectangle A?

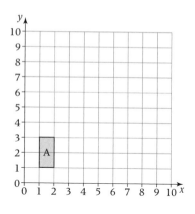

  d Write the area of   **i** rectangle A   **ii** rectangle B.

  e How many times larger than the area of rectangle A is the area of rectangle B? Write this value as a power of 3.

2 Copy this diagram.

  a Enlarge triangle X by scale factor 2, centre (1, 2). Label the image Y.

  b Write the area of

    i triangle X

    ii triangle Y.

  c How many times larger than the area of triangle X is the area of triangle Y? Write this value as a power of 2.

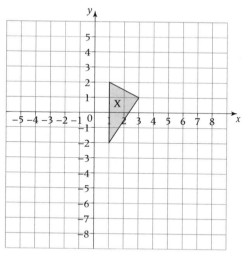

  d Enlarge triangle X by scale factor $\frac{1}{2}$, centre (−3, 2). Label the image Z.

  e Write the area of   **i** triangle X   **ii** triangle Z.

  f How many times larger than the area of triangle X is the area of triangle Z? Write this value as a power of $\frac{1}{2}$.

1 Describe fully the single transformation that maps

  a triangle A onto triangle B

  b triangle B onto triangle A.

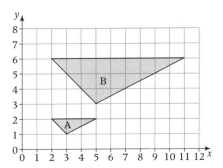

2 Describe fully the single transformation that maps

  a trapezium X onto trapezium Y

  b trapezium Y onto trapezium X.

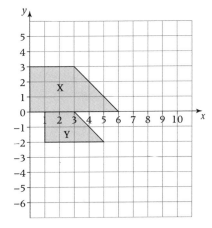

3 These trapeziums are similar shapes.

  a Find the missing lengths marked x and y.

  b Find the area of each trapezium.

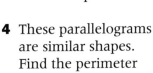

10 cm

5 cm

x

10.8 cm

9 cm

y

4 These parallelograms are similar shapes. Find the perimeter of the larger parallelogram.

3 m

5 m

14 m

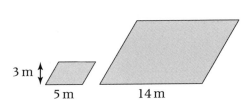

1 **a** Work out the length of PS.
   b Work out the length of RQ.

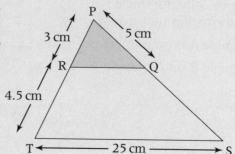

2 Betty brings back two boxes of 'Mackay' chocolates from Scotland. Each box has the same picture on the front.

The picture on the 320 g box is 102 mm by 170 mm.

The picture on the 480 g box is 140 mm by 226 mm.

Show that the two rectangles are *not* mathematically similar.

3 Copy this diagram, and extend both axes up to 8.
Enlarge the shaded kite by scale factor 2, centre O.

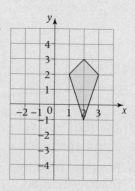

4 Describe fully the transformation that maps the shaded rectangle onto the *un*shaded rectangle.

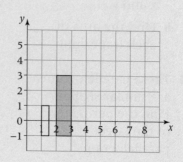

1 a Simplify these ratios.

    i $3\,m : 3000\,m$ **ii** $25p : £3$ **iii** $30 : 12 : 18$ **iv** $\frac{1}{4} : \frac{2}{3}$

  b Find the missing values in these equivalent ratios.

    i $a : 3 = 8 : 12$ **ii** $b : 5 = 1 : 3$ **iii** $7 : x = 8 : 5$

2 Solve these double-sided equations.

  a $3(x + 2) = x + 8$ **b** $5(x - 3) = 3(x - 1)$

  c $4(2 - p) = 3(p + 5)$ **d** $5(3 - y) = 3(4 - y)$

  e $4a + 2(a - 5) = 8$ **f** $(t + 2)(t + 3) = (t + 4)^2$

3 These two parallelograms are mathematically similar.

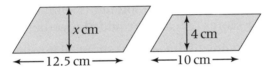

Find the missing value x and hence the area of the larger parallelogram.

4 A teacher marks a class set of books and records the number of minutes that each individual book takes him.

No. of mins, m	$0 < m \leqslant 1$	$1 < m \leqslant 2$	$2 < m \leqslant 3$	$3 < m \leqslant 4$	$4 < m \leqslant 5$
Frequency	1	6	10	7	6

  a Draw a cumulative frequency table for these data.

  b Draw a cumulative frequency diagram for these data.

  c Estimate (from the graph)

    i the median

    ii the interquartile range.

1 A bag contains red, blue and yellow balls. The probability of selecting a red ball at random from the bag is $\frac{3}{8}$.

   a If a ball is selected from the bag, what is the probability that the ball is not red?

   b If there are 40 balls in the bag, how many of them are blue or yellow?

2 A letter is selected, at random, from the letters of the word 'INDEPENDENCE'. Find the probability that the letter chosen is

   a a D   **b** a consonant

   c not an E   **d** from the second half of the alphabet.

3 A spinner has equally sized sections numbered 1 to 4. A fair dice is numbered 1 to 6. The spinner is spun and the dice is thrown. The scores on each are added together.

   a Copy and complete the table to show all possible outcomes.

		Dice					
		1	**2**	**3**	**4**	**5**	**6**
Spinner	**1**	2	3	4			
	2	3					
	3						
	4						

   b Find the probability that the sum of the scores is

   i 10   **ii** 7   **iii** not 5   **iv** 2 or 3.

4 A bag contains 5 aquamarine, 3 amethyst and 2 topaz gemstones. A fair coin is thrown and a gemstone selected from the bag at random. Find the probability of selecting

   a a head and an aquamarine gemstone
   b a tail and a topaz gemstone.

   c What can you say about the events 'select an amethyst' and 'throw a head'? Explain your answer.

1 A jewellery box contains 10 rings, all identical except for the precious stone used in each. 7 of the rings contain a diamond and 3 contain a ruby. A ring is selected from the box at random and replaced in the box. A second ring is then selected at random.

  a Copy and complete the tree diagram to show all possible outcomes.

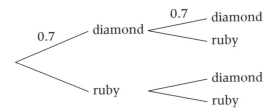

  b Find the probability of selecting
   i two diamond rings
   ii two ruby rings
   iii a ruby then a diamond ring
   iv one of each stone in any order.

2 A bag contains 6 magenta counters and 4 turquoise counters. A counter is selected from the bag at random and *not* replaced. A second counter is then selected at random.

  a Copy and complete the tree diagram to show all possible outcomes.

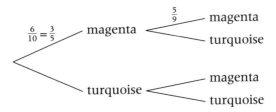

  b Find the probability of selecting
   i two magenta counters
   ii two turquoise counters
   iii a magenta then a turquoise counter
   iv one of each counter in any order.

111

1 Three coins are thrown together.

  a List all the possible outcomes of this experiment.

  b Find the probability that the three coins show

    i three heads

    ii a head and two tails in any order.

2 Daisy carries out a statistical experiment. She throws a coin 500 times. She throws a head 400 times.

  a Is the coin fair? Explain your answer.

  Daisy then throws two fair coins once.

  b Copy and complete the probability tree diagram to show the outcomes.

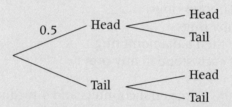

  c Find the probability of throwing a head and a tail in any order on this throw of Daisy's fair coins.

3 A teacher requires two classroom monitors to be in charge of the register and collecting in work. Two students are to be chosen at random from a class of 25 students: 15 boys and 10 girls.

  a Draw a tree diagram to show all possible outcomes.

  b Find the probability that the two students chosen are both boys.

  Hint: Choosing two students is the same as choosing without replacement. Remember to reflect this in the fractions that you use.

1 a Divide £250 in the ratio 2 : 3.

   b If y is proportional to x and when $x = 4$, $y = 6$, find

   i   a formula connecting x and y

   ii the value of y when $x = 5$

   iii the value of x when $y = 12$.

2 Copy and complete this table to solve

$$x^3 - x = 70$$

to 2 decimal places.

x	$x^3 - x$	Comment
4.2	69.888	Too small
4.3	75.207	Too big

Hint: Look at 3 decimal places in order to round to 2 decimal places.

3 a Calculate the volume of a cube whose edges are all 5 cm.

   b Calculate the volume of a cube whose surface area is 384 cm².

4 A coin and a dice are thrown together.

   a Draw a table to show all the possible outcomes.

   b Find the probability of throwing

   i   a head on the coin and a 3 on the dice

   ii a tail on the coin and an even number on the dice.

1 Aimée leaves her home in Clutton and cycles to Chester, 20 km away. On the outward journey, Aimée cycles at an average speed of 20 km/h but stops halfway for a 15 minute break. On reaching Chester, she spends 45 minutes at a friend's house trying to fix a flat tyre on her bicycle. She leaves her bike there in Chester and catches a bus back home. She arrives home in Clutton 25 minutes later.

  a Construct a distance–time graph to represent Aimée's journey.

  b How far did Aimée cycle before her break on the outward journey?

  c Work out the average speed of the bus in

    i km/h   **ii** miles per hour.

  Hint: 1 km ≈ 0.62 miles

2 Finlay sets out at 1 p.m. from Farndon and walks at a speed of 6 km/h to Malpas, 12 km away.
At 1.30 p.m. Joe jumps on his bike in Malpas and cycles to Farndon, reaching his destination in half an hour.

  a Construct a distance–time graph to show both journeys.

  b Work out the average speed at which Joe cycles.

  c Use your graph to work out when the two boys meet.

3 Sketch a graph of depth against time for this container, if water is poured in at a steady rate.

4 This is a graph of depth against time, if water is poured into a container at a steady rate.

Sketch the container.

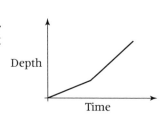

1 This is a conversion graph to convert British pounds to Australian dollars.
Use the graph to convert

   a **i** £50 to Australian dollars
   ii AU$96 to pounds.

   b If Julien has £60 and Alison has AU$120, who can purchase the most British goods if Alison converts her money?

   c Write the equation of the line, stating the meaning of any letters used.

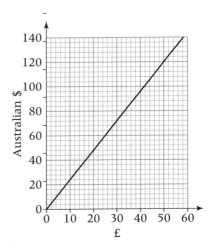

2 Plot a graph to represent the total cost, of hiring a plumber if his call out charge is £30 and he charges £15 an hour labour. The longest that he will take is 8 hours.

   a Use your graph to work out the cost of hiring a plumber for 4 hours.

   b How long did he spend on a job if the bill was £112.50?

   c Write the equation of the line, stating the meaning of any letters used.

3 A sales company reimburses its staff for travel expenses. This graph shows the amount reimbursed per km travelled.

Find the equation of the line, stating the meaning of any letters used. Interpret the meaning of m and c and discuss the limitations of this scheme for reimbursing travel costs.

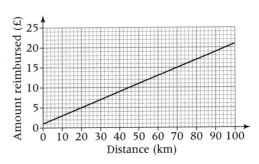

1 Which graph best represents each of these statements?

  a The price of the US dollar against the pound had begun to rise but is now falling steadily.

  b The cost of a company's shares has risen steadily during the past three months.

  c The cost of petrol, which had been rising steadily in the last year, is now rising rapidly.

  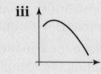

2 Lauren leaves home and drives 70 miles to see her Grandma. This travel graph shows her journey.

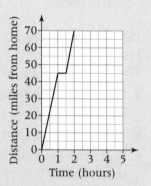

  a Explain what happened after 1 hour.
  Lauren stays for an hour and then drives home at a steady speed. Her return journey takes $1\frac{1}{2}$ hours.

  b Copy and complete the graph to show this journey.

  c What was the average speed for Lauren's journey home?

3 The diagram shows a rectangle x metres by $5 - x$ metres.

  a Write an expression for the area of the rectangle.

  b Let the area of the rectangle be y m². Plot a graph of x against y.

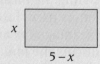

  c Use your graph to find
    i the maximum possible area of the rectangle
    ii the dimensions of the rectangle if the area is 4 m².

1 In the summer sale, all clothes have a 20% discount.

  a What is the sale price of these items?

    i a coat costing £120 full price

    ii a pair of trousers costing £45.50 full price

  b Paula buys a selection of clothes at sale prices. Work out the original cost of each of the items that she purchased.

    i a skirt at a sale price of £40

    ii a jumper at a sale price of £31.96

    iii a jacket at a sale price of £71.99

2 **a** Copy and complete this table to generate three coordinates for the line graph $y = 5 - 2x$.

x	0	1	2
y			1

  b Draw an x-axis from −1 to 3 and a y-axis from −3 to 5. Plot the graph of $y = 5 - 2x$ on your axes.

  c Plot the graph of $y = 4x - 4$ on the same set of axes.

  d Use your graphs to solve the simultaneous equations

$$y = 5 - 2x$$
$$y = 4x - 4$$

3 If a, b, c, r and h represent lengths, write each of these expressions that represent an area.

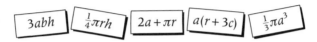

$$3abh \quad \tfrac{1}{4}\pi rh \quad 2a + \pi r \quad a(r + 3c) \quad \tfrac{1}{3}\pi a^3$$

4 Change these measurements to m².

  a 4 km²   **b** 2500 cm²   **c** 80 000 mm²   **d** 0.055 km²

1 Find the missing sides in each of these right-angled triangles, giving your answers to 3 significant figures.

a
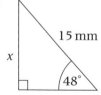
15 mm

x

48°

b
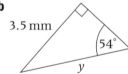
3.5 mm

54°

y

c
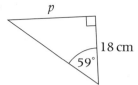
p

18 cm

59°

d
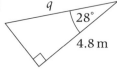
q

28°

4.8 m

e
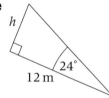
h

24°

12 m

f
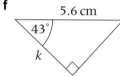
5.6 cm

43°

k

2 A ladder rests against a wall.

The distance from the base of the wall to the top of the ladder is 7.7 m.

The ladder makes an angle of 50° with the ground.

How long is the ladder?

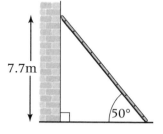

7.7 m

50°

3 A rocky outcrop, R, is 55 km due west of a lighthouse, L, and due north of a ship, S. The lighthouse is on a bearing of 038° from the ship.

How far is the ship from the rocks, to the nearest km?

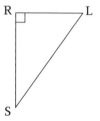

R   L

S

1 Find the missing angles in each of these right-angled triangles, giving your answers to 3 significant figures.

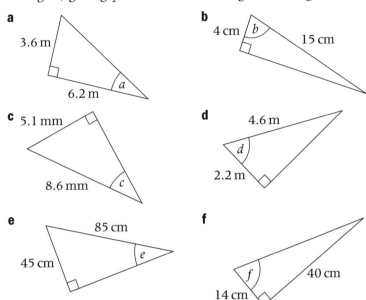

a
3.6 m
6.2 m
a

b
4 cm
b
15 cm

c 5.1 mm
8.6 mm
c

d
4.6 m
d
2.2 m

e
85 cm
45 cm
e

f
f
40 cm
14 cm

2 A rocky outcrop, R, is 45 km due north of a ship, S, and 30 km due east of a lighthouse, L. On what bearing is the lighthouse from the ship?

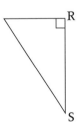

Hint: Find the angle LSR using trigonometry but remember that the bearing of L from S is measured clockwise from north.

3 Sandeep lies on the ground 15.5 m from a tree. By knowing this distance and measuring the angle of elevation to the tree, she is able to work out that the tree is 6.3 m tall. What was the angle of elevation that she used?

Hint: Sketch a diagram and mark on the two distances given. Use trigonometry to find the angle.

1 ABCD is a parallelogram.
AB = CD = 12.4 cm, AD = BC = 8.2 cm. Angle BAD = 112°.

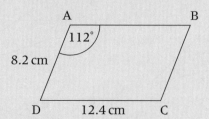

Work out the area of the parallelogram.

Hint: Find the vertical height using trigonometry.

2 A chord AB is drawn inside a circle,
centre O, diameter 10 cm.

The angle AOB = 47°.

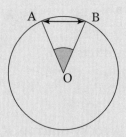

Find the length of AB correct to
1 decimal place.

3 PQRS is a triangle with PS = 12.2 cm, QR = 9.5 cm
and angle PQS = 57°.

Find θ.

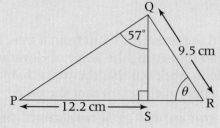

N1 HW1

1. a $\frac{7}{8}$ b $\frac{11}{35}$ c $2\frac{37}{56}$ d $1\frac{233}{240}$
   e $\frac{1}{3}$ f $4\frac{1}{2}$ g $6\frac{2}{3}$ h $5\frac{1}{3}$

2. a $x = 4$ b $x = \frac{2}{3}$ c $x = 4$
   d $x = -2\frac{1}{2}$ e $x = 10$ f $x = 1\frac{3}{4}$

3. 5.83 cm

4. a 0.45 b 0.85 c 8

S1 HW1

1. a 3.6 b 12.4 c 5.3
   d 1.0 e 101.0 f 1032.9

2. a 63 b 0.492 c 3.6
   d 0.00853 e 0.086 f 0.1002

3. a 3 b 60 c −0.03
   d 37 e 24 f −70
   g −17 h −15

4. a $36 = 2 \times 2 \times 3 \times 3$ b 18 c 180

5. a Prism b Prism c Not prism
   d Not prism e Not prism

A1 HW1

1. a 4 b 7 c 30 d −6 e −30
   f −6 g −12 h −5 i $-\frac{1}{2}$

2. ai $60 = 2 \times 2 \times 3 \times 5$
      $108 = 2 \times 2 \times 3 \times 3 \times 3$
   aii 12 bi 326.6 bii 0.3266

3. 4350 cm³

4.

Height h (cm)	Frequency
$110 < h \leqslant 115$	4
$115 < h \leqslant 120$	6
$120 < h \leqslant 125$	11
$125 < h \leqslant 130$	7
$130 < h \leqslant 135$	2

N2 HW1

1. a 270 b 3.4 c 7.89
   d 50 e 9.84 f 1.013
   g 42 050 h 0.0031

2. a $12xy$ b $8p^2$ c $3a^3$
   d $8ab + bc$ e $5m + 3n$ f $3x^3 - x^2$
   g $\frac{2g}{k}$ h $8y$

3. 154 cm²

4. a 4 b 5
   c No rotational symmetry d 2

A2 HW1

1. a 5052 b 833 c 8188
   d 253 e 17 444 f 2451
   g 305 h 86 142

2. a $5x + 20$ b $p^2 - 8p$
   c $3a - 3b + 6$ d $2h - 24$
   e $x^2 + 12x + 27$ f $y^2 - 4y + 4$
   g $2a^2 - 7a - 15$ h $2p^2 - 2q^2$

3. 44.0 cm² (1 d.p.)

4. ai 7.36 (2 d.p.) aii 8
   aiii 2 aiv 12
   b The mode in this case does not
     accurately reflect the spread of
     the data; it does not take into
     account all values in the data set.

D1 HW1

1. a 0.586, 0.856, 5.68, 5.86, 6.85,
     58.6, 85.6, 685
   b 0.014, 0.12, 0.124, 0.142, 0.41,
     1.42, 2.1, 4.12

2. a 2 b −11 c 12 d −1
   e 12 f −1 g −16 h 16

3. a 15 cm b 18.3 cm (1 d.p.)

4. 4 cm

N3 HW1

1. a 32 b 130 c 0.9
   d 393 e 0.027 f 8500
   g 6.68 h 60 000

2. a $y = -4$ b $x = 1$

3. 1964 cm² (4 s.f.)

4. a 75 b 52% c 55.6% (1 d.p.)

S2 HW1

1. $\frac{3}{4}$, 0.75, 75%; $\frac{2}{5}$, 0.4, 40%;
   $\frac{7}{20}$, 0.35, 35%; $\frac{1}{20}$, 0.05, 5%;
   $\frac{5}{8}$, 0.625, 62.5%

2. a $x > 2$ b $x \leqslant 6$ c $n \geqslant -5$
   d $x \leqslant -6$ e $x \geqslant 12$ f $x < 14$
   g $3\frac{1}{3} < x < 3\frac{1}{2}$ h $-5 \leqslant x \leqslant 5$

3 i 153.9 m^2 (1 d.p.)
  ii 44.0 m (1 d.p.)
4 a Similar opinions may be obtained from similar people.
  b Use the school register and choose every nth person (say, for $n = 8$).
  c

	Yes	No
Boys		
Girls		

A3 HW1

1 ai $\frac{7}{9}$ **aii** $\frac{49}{99}$ **aiii** $\frac{73}{99}$ **aiv** $\frac{345}{999} = \frac{115}{333}$
  b $0.\dot{4}$ **c** $0.\dot{8}\dot{3}$ **d** $0.\dot{1}42\,85\dot{7}$
2 a $x = 35$ **b** $x = 4$ **c** $x = 2$ **d** $x = 36$
3 a 45 cm^2 **b** 89.1 m^2 (1 d.p.)
4 77.6% (3 s.f.)

D2 HW1

1 a 0.625 **b** $0.1\dot{6}$ **c** 0.0625
  d $0.\dot{8}$ **e** $0.41\dot{6}$
2 Sequence $1 \rightarrow 5 - n$
  Sequence $2 \rightarrow 3n + 1$
  Sequence $3 \rightarrow (n + 1)^2$
  Sequence $4 \rightarrow 4n$
  Sequence $5 \rightarrow 4n^2$
3 a $a = 63°$ $b = 63°$ $c = 54°$ $d = 126°$
  b $e = 71°$ $f = 64°$ $g = 45°$
4 14.5

A4 HW1

1 a $\frac{3}{4}$ **b** $\frac{1}{2}$ **c** $\frac{7}{12}$ **d** $\frac{14}{15}$
  e $\frac{17}{36}$ **f** $\frac{53}{56}$ **g** $\frac{2}{15}$ **h** $1\frac{1}{6}$
2 a $x = 5$ **b** $q = -2$
  c $a = 1$ **d** $y = -6$
3 $\frac{30}{x+2} = 5 \rightarrow x = 4$
4 $9 \times 140° = 1260°$
5

3	1 5 8
4	0 2 3 8
5	1 2 3 4 5 6
6	2 2 4 5 6
7	0 0 1 2 4 7 9
8	0 2 4 7 9

Median mark = 63
Key: $3 \mid 1 = 31$

D3 HW1

1 a 11.48 **b** 8.78 **c** 4.116 **d** 4.09
2 ai $(0, 3)$ **aii** 1
  bi $(0, 5)$ **bii** 4
  ci $(0, 8)$ **cii** -1
  di $(0, 2)$ **dii** 3
  ei $(0, \frac{3}{2})$ **eii** $-\frac{1}{2}$
  fi $(0, -\frac{5}{3})$ **fii** 3
3 a 1152 m^2 **b** 1187.5 cm^2 (1 d.p.)
4 a

Negative correlation – as age of car increases, price decreases
  b £7100
  c Other factors such as condition, mileage, service history etc have not been included, hence not all that reliable. Line of best fit drawn by sight – may not be entirely accurate.

N4 HW1

1 a 3753 **b** 375 300 **c** 0.695
  d 3.753 **e** 6.95
2 $c = 3$
3 a $x = 106°$ since the angle at the centre is twice the angle subtended at the circumference ($212°$)
  $y = 74°$ since the angle at the centre is twice the angle subtended at the circumference ($148°$)
  b $z = 32°$ since the angle in a semicircle is a right angle

4 a

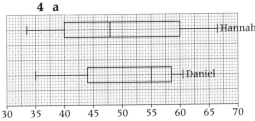

b Hannah has a symmetrical distribution. Daniel's is negatively skewed. Daniel's distribution has a smaller range i.e. he is more consistent.

S3 HW1

1 a $\frac{1}{20}$, 0.12, 23%, 0.45, $\frac{4}{5}$, $\frac{41}{50}$, 1

  b 2%, $\frac{17}{100}$, 0.24, 0.3, $\frac{1}{2}$, 65%, $\frac{39}{50}$, 90%

2 ai 17, 15, 13, 11, 9   **aii** $n = 15$

  b $u_n = 4n - 1$

3 a 45°   **b** 135°

4 a Add in option boxes

  b In your opinion, what is the maximum number of hours a day that a child should watch television?

    Less than 1 hour

    1–2 hours

    More than 2 hours

N5 HW1

1 a 3.3 kg   **b** 19.8 kg

  c 33 kg   **d** 79.2 kg

2 a $y = 2x + 1$ **b** $y = 3x - 6$

  c $y = 3 - x$

3 a Flag shape with vertices at (1, 2), (4, 2), (4, 1), (3, 1), (3, 2) labeled G

  b Flag shape with vertices at (1, 4), (4, 4), (4, 5), (3, 5), (3, 4) labeled H

  c Translation column vector $\begin{pmatrix} 0 \\ 6 \end{pmatrix}$

4 a Cannot both occur at the same time; 2 is not an odd number

  bi $\frac{1}{2}$ **bii** $\frac{1}{6}$ **biii** $\frac{2}{3}$ **biv** $\frac{1}{3}$

A5 HW1

1 ai 54 **aii** 144 **aiii** 96 **aiv** 198

  bi 68%   **bii** 95%

  biii 56.5%   **biv** $46\frac{2}{3}$%

2 ai $u_n = 2n + 7$ **aii** $u_n = 20 - 3n$

  aiii $u_n = \frac{n}{4} + 3$ **b** $u_n = n^2 - 3n$

3 a Flag with vertices at (1, −1), (4, −1), (3, −2), (3, −1) labeled B

  b Flag with vertices at (−1, −1), (−4, −1), (−3, −2), (−3, −1) labeled C

  c Reflection in the line $y = -x$

4

Starter	Main
Soup	Lasagne
Soup	Rack of lamb
Soup	Roast chicken
Soup	Beef en croute
Paté	Lasagne
Paté	Rack of lamb
Paté	Roast chicken
Paté	Beef en croute
Brioche	Lasagne
Brioche	Rack of lamb
Brioche	Roast chicken
Brioche	Beef en croute

S4 HW1

1 a 0.12 **b** 0.03 **c** 0.84 **d** 3.1

  e 15 **f** 0.525 **g** 36.5 **h** 230

2 a $y = 3x + 4$ **b** $y = 3 - 2x$

3 a $a = b = 77°$ **b** $x = 60°$ $y = 120°$

4 a

  b 66

N6 HW1

1 a 1   **b** 500   **c** 256   **d** 7
   e 6   **f** $\frac{1}{5}$ **g** $\frac{1}{2}$ **h** 8

2 a 7, 9, 11, 13, 15
   b 11, 10, 9, 8, 7
   c 6, 11, 18, 27, 38
   d 2, 9, 28, 65, 126
   e $\frac{1}{2}, \frac{1}{6}, \frac{1}{12}, \frac{1}{20}, \frac{1}{30}$

3 a 15.8 cm (3 s.f.)   **b** 10 mm

4 a

0	8 9
1	2 5 7
2	1 5 6 7 8
3	0 1 2 3 6 9
4	0 1 4 5 5 Key: 0	9 = 9 s

   b Negatively skewed
   ci 30   **cii** 19

A6 HW1

1 a 28   **b** 10   **c** 15   **d** 55
   e 48   **f** 37   **g** 2   **h** 13

2 ai (2, 5)   **aii** (3, −4)   **aiii** $(\frac{2}{3}, -1)$
   b $y = 3 - 2x$

x	−2	0	2
y	7	3	−1

3 a 4.12 units (3 s.f.)
   b 6.71 units (3 s.f.)
   c 5.39 units (3 s.f.)

4 a $\frac{3}{10}$ **b** 240

D4 HW1

1 a $8 + 7 \times 3 = 29$ **b** $12 \div (5 - 1) = 3$
   c $3 \times (9 - 2) \times 4 = 84$
   d $12^2 \times (5 - 3) = 288$
   e $(24 \div 4 - 2)^{\frac{1}{2}} = 2$

2 a $x = -3$ or -7 **b** $x = 5$ or -9
   c $x = 9$ or -4 **d** $y = 0$ or 5
   e $t = 0$ or -4

3 a Congruent AAS
   b Not congruent, only similar

4 a

	Red	Blue	Yellow	Total
Girls	5	4	5	14
Boys	6	8	2	16
Total	11	12	7	30

   bi $\frac{2}{5}$ **bii** $\frac{1}{5}$ **c** 49

S5 HW1

1 a $2\sqrt{3}$ **b** 5   **c** $5\sqrt{2}$
   d $2\sqrt{5}$ **e** 9   **f** $27 + 10\sqrt{2}$

2 a $x = b - a$ **b** $x = \frac{z}{y}$
   c $x = \frac{r^2 - q}{p}$ **d** $x = (\frac{q}{y - p})^{\frac{1}{2}}$
   e $x = t - w$ **f** $x = abc$
   g $x = \frac{a}{p - q}$ **h** $x = h - \frac{m}{r}$

3 a 200 cm^2 **b** 14.1 cm (3 s.f.)

4 a 2.28   **b** 1

N7 HW1

1 a 5^6 **b** 3^3 **c** a^{11} **d** $8^0 = 1$
   e b^{-5} **f** $4^1 = 4$ **g** 7^{12} **h** 2^{11}

2 $u_n = 5n + 7$

3 a   **b** 290°
   c 110°, 30°, 40°

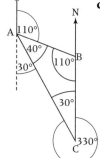

4 a

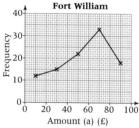

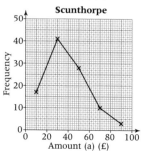

b Modal class for Fort William is $60 \leqslant a < 80$ and for Scunthorpe is $20 \leqslant a < 40$, where a is the amount spent. Modal amount spent in Fort William is higher, perhaps due to Ben Nevis being close by etc.

D5 HW1

1 $3\sqrt{3}$

2 a $p = 3$, $q = 2$ **b** $a = 2$, $b = -1$
   c $x = 5$, $y = 1$

3 b Isosceles   **c** 4.58 cm (3 s.f.)

4 a

Mass, m, kg	Cumulative frequency
$2 < m \leqslant 2.5$	1
$2.5 < m \leqslant 3$	4
$3 < m \leqslant 3.5$	7
$3.5 < m \leqslant 4$	10
$4 < m \leqslant 4.5$	3

b

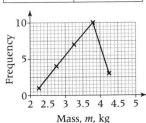

S6 HW1

1 a £160 800   **b** 130%

2 a $x = -3$ or -4 **b** $x = -2$ or -7
   c $x = 1$ or -5 **d** $x = 4$ or 10
   e $x = 8$ or -3 **f** $x = 0$ or 9

3

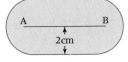

4 a

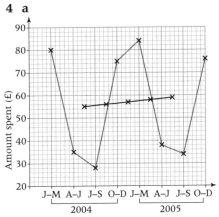

b 54.5, 55.5, 56.25, 57.75, 58
c Increasing

A7 HW1

1 373 g cod, 250 g wild Alaskan salmon, 625 g mashed potatoes, 3 eggs, 200 g breadcrumbs

2 a $x = 5$, $y = 1$ **b** $x = 2$, $y = -1$
   c $x = 4$, $y = 2$

4 a $7.5 \rightarrow 22.5$; $12.5 \rightarrow 100$;
   $17.5 \rightarrow 210$; $22.5 \rightarrow 112.5$;
   $27.5 \rightarrow 55$; mean $= 16\frac{2}{3}$
   bi $15 \leqslant t < 20$ **bii** $15 \leqslant t < 20$

N8 HW1

1 Lower bound is 11.25 cm^2, upper bound is 19.16 cm^2

2 a

x	-2	-1	0	1	2	3	4
x^2	4	1	0	1	4	9	16
$-2x$	4	2	0	-2	-4	-6	-8
3	3	3	3	3	3	3	3
y	5	0	-3	-4	-3	0	5

bi $x = 3$ or -1 **bii** $x = 0$ or 2
biii $x = 3.24$ (2 d.p.) or -1.24
   (2 d.p.)
biv $x = 4$ or -1

3

4 a $6 < t \leqslant 9$ **b** 8 hours

S7 HW1

1 a LB = 11.5 m   UB = 12.5 m
  b LB = 14.75 s   UB = 14.85 s
  c LB = 1150 g   UB = 1250 g
  d LB = 24.45 kg UB = 24.55 kg
  e LB = 3.045 l   UB = 3.055 l
  f LB = 4.385 m UB = 4.395 m

2 a

x	-2	-1	0	1	2	3	4
x^2	4	1	0	1	4	9	16
$-x$	2	1	0	-1	-2	-3	-4
-4	-4	-4	-4	-4	-4	-4	-4
y	2	-2	-4	-4	-2	2	8

b 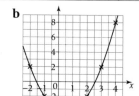   **c** $\left(\frac{1}{2}, -4\frac{1}{4}\right)$

3 1800 cm³

4 0.35

D6 HW1

1 ai 1 : 1000   **aii** 1 : 12
  aiii 5 : 2 : 3   **aiv** 3 : 8
  bi $a = 2$ **bii** $b = 1\frac{2}{3}$
  biii $x = 4\frac{3}{8}$

2 a $x = 1$ **b** $x = 6$ **c** $p = -1$
  d $y = 1.5$ **e** $a = 3$ **f** $t = -3\frac{1}{3}$

3 $x = 5$ cm Area = 62.5 cm²

4 a

Number of mins, m	Cumulative frequency
$\leqslant 1$	1
$\leqslant 2$	7
$\leqslant 3$	17
$\leqslant 4$	24
$\leqslant 5$	30

b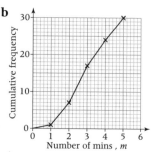

ci 2.7 mins ≈ 2 mins 40 secs
cii 1.7 mins ≈ 1 min 40 secs

A8 HW1

1 a £100 and £150
  bi $y = \frac{3}{2}x$ **bii** 7.5 **biii** 8

2 4.20 (2 d.p.)

3 a 125 cm³ **b** 512 cm³

4 a

	1	2	3	4	5	6
H						
T						

bi $\frac{1}{12}$ **bii** $\frac{1}{4}$

S8 HW1

1 ai £96   **aii** £36.40
  bi £50   **bii** £39.95
  biii £89.99

2 a

x	0	1	2
y	5	3	1

b, c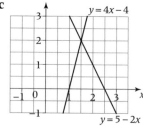

d $x = 1.5$, $y = 2$

3 The expressions that represent area are $\frac{1}{4}\pi rh$ and $a(r + 3c)$.

4 a 4 000 000 m² **b** 0.25 m²
  c 0.08 m²   **d** 55 000 m²